*economic growth
and
environmental decay*

the solution becomes the problem

PAUL W. BARKLEY
Washington State University

DAVID W. SECKLER
Colorado State University

economic growth and environmental decay

the solution becomes the problem

 Harcourt Brace Jovanovich, Inc.
New York / Chicago / San Francisco / Atlanta

THE HARBRACE SERIES IN BUSINESS AND ECONOMICS

PHOTOGRAPH ACKNOWLEDGMENTS

Page 1: Harbrace; 22: Harbrace; 34: Harbrace;
50: Harbrace; 62: Harbrace; 84: Harbrace;
98: Richard Conrat from Photofind; 124:
Harbrace; 146: Moulin Studios and Save-the-
Redwoods League; 156: Gerhard Gscheidle;
172: UPI; 184: Harbrace. Cover photo:
Harbrace.

Library of Congress Catalog Card Number:
74-172200

Printed in the United States of America
ISBN: 0-15-518795-3

THIS BOOK WAS PRINTED ON RECYCLED
PAPER

To Karen and Kurt, Daniel, Christian, and Andrew

preface

The groundwork for this book was laid in 1966 when we were colleagues in the Department of Economics at Colorado State University. At that time, environmental quality was just emerging as a major theme in the literature of economics. Its emergence raised serious questions about the nature of economic theory and economic policy. Our discussions of these questions eventually led to our decision to write a small book questioning the emphasis that contemporary economics places on the process of growth.

Two major themes are presented in this book. The first deals with economic growth and its effect on society and, more importantly, on environment. Economic growth has, in the past, contributed much to man's survival, but now, in the closing years of the twentieth century, some second thoughts about growth may be in order. In addition, greater attention must be paid to the quality of the environment. The second theme deals with the relationship between economics and the market system, and the environment. We hope to show how an economic system based on private decision-making processes leaves much opportunity for environmental decay — not necessarily through conscious efforts of individuals — but through gaps in ownership patterns, lapses in incentives, and uncertainties regarding the future. Correcting these ills will not be easy, and the prescriptions found herein are, perhaps, only stopgap measures. The fact that the issues are carefully laid out for a reader's examination is tremendously important, however.

Occasionally, a reader will come upon a reference to attempts (usually made by governments) at decision-making in a nonprice world. Discussions of benefit/cost analysis and cost/effectiveness show how governments have tried to fill gaps left by an imperfect organization of economic activity. Although the attempts have been valiant, the gaps remain; the efforts to close them may have actually contributed to the gravity of the current economic and environmental problems.

Writing this book has been enjoyable for us. It has forced us to think through some things that until now have gone undefined in our professional work; it has forced us to bring into focus the rather harsh interface between economics and environmental problems; it has forced us to wonder if a new field of endeavor — *economic ecology* — may not emerge in the next few years. We think our readers will be similarly stimulated.

We are indebted to many for their contributions to this book. In terms of intellectual tradition, we have borrowed heavily from the original and nearly forgotten thinking of the Classical school of economists, from Adam Smith through Thomas Robert Malthus to John Stuart Mill. Our thinking has been influenced (some may even say warped) by the great institutionalist Thorstein Veblen. Also, readers will recognize the pervasive influence of such contemporary economists as John Kenneth Galbraith, Kenneth Boulding, and Ezra Mishan.

We take this opportunity to thank our many reviewers — Will Siri, Andrew Schmitz, Richard B. Norgaard, Ezra Sadan, Martin Curry, Dirck Ditwiler, LeRoy Rogers, Ralph Loomis, Earl Bell, and Thomas Palm. We are also grateful to William J. Baumol for valuable suggestions and encouragement.

The secretarial staffs in the Departments of Agricultural Economics at the University of California at Berkeley and Washington State University have our undying thanks, as do the administrators of these departments who offered continued support and encouragement.

Finally, we owe a special debt of gratitude to our good friend and editor, Don DeLaura. He not only put the prose in order but contributed to the book's message.

Paul W. Barkley
David W. Seckler

PART ONE

1 | **_chapter one_**
The Solution Becomes The Problem

10 | **_chapter two_**
Economic Growth And The Environment

22 | **_chapter three_**
The Quantity Of Product–The
Quality Of Life

34 | **_chapter four_**
The Meaning Of Economic Growth

PART TWO

50 | **_chapter five_**
Preferences And Values

62 | **_chapter six_**
Two Tools Of Economic Analysis–
Supply And Demand

84 | **_chapter seven_**
A Nonmarket Tool–Benefit/Cost Analysis

98 | **_chapter eight_**
Market Failure–Externalities

124   ***chapter nine***
Market Failure–Collective Goods

146   ***chapter ten***
The Logic Of Conservation

PART THREE

156   ***chapter eleven***
Recreation And Cost/Effectiveness–
Some Further Applications

172   ***chapter twelve***
On The Strategy And Tactics
Of Environmental Control

184   ***chapter thirteen***
Economic Growth And Environmental Decay

part one

chapter one

The Solution Becomes The Problem

Human history is largely written in terms of the struggle between man and nature over the terms of man's existence. The struggle has been won, at least for a time, in the developed nations of the world, and these nations have become affluent societies.

But they are affluent only in material and pecuniary respects. Economic progress has not come without costs. As man's will has subdued nature, many natural amenities have been destroyed. Clean air and water, open space, trees, and wildlife have become scarce as natural phenomena have been transformed into man-made commodities. It is now clear that the process of economic growth has been a centuries-long process of substituting man-made goods for natural amenities. When people were poor and amenities abundant, this substitution made sense. In those parts of the world where people are now wealthy and amenities scarce, it may no longer make sense. Serious thinkers

1

are beginning to realize that man's interest in economic growth may conflict with his interest in maintaining a high-quality environment. Economic growth — as it is popularly understood — requires large amounts of land, forests, water, and other natural resources. As these resources are consumed to produce houses, automobiles and televisions, they no longer exist as amenities. The time has come when serious choices must be made between growth and quality. This book is about those choices.

Like most recent books about environmental problems, this one does not offer specific answers to questions regarding what courses of action the nation should take. But unlike so many others, it does not dwell on the definition of the problem nor on the recounting of ecological horror stories. It is intended to provide a framework within which the growth versus environment question can be viewed. It examines how the advanced nations got where they are and what might be done to forestall continued deterioration of their environments.

Economics is a reasonable place to start. In its most honored modern formulation, this discipline is defined as the one that studies choices made among alternative and competing ends. The definition describes the dilemma: The alternative and competing goals are a quality environment and continued economic growth. Economists have built an elaborate analytical framework useful in studying difficult choices. Because of the importance of *choice* to the subject of this book, a digression into the philosophic problems of choice seems warranted.

The Greek philosophers created the foundations of modern science and humanism. While they were doing this, however, their society was just emerging from a long era of paganism in which man was considered hardly more than a plaything of the gods. During this emergence, strong forces were encouraging man to think of himself as a heroic figure capable of determining his own fate. Equal and opposite forces were insisting that he was no more than a tragic dupe in the larger scheme of things.[1]

Nowhere is this Greek dilemma better illustrated than in the tragedy, *Oedipus Rex*. As a young prince, Oedipus was entrusted to the care of a peasant family to protect him from his father. The father had been told that he would be slain by his son and, therefore, wished to kill the son first. But when Oedipus was a young man, an oracle told him that he was destined to kill his father and marry his mother. Thinking that the peasant couple was his true father and mother, Oedipus fled their home in order to avoid this fate. He became a highwayman, and after much wandering, he and his band happened on a traveling king whom Oedipus killed. He married the widowed queen and took the kingdom. The king was his father, the queen his mother. When he later discovered this, he renounced his throne and, having blinded himself, spent the rest of his life a pitiful beggar wandering in his former kingdom.

The fascination of *Oedipus Rex* is its tantalizing *illusion* of choice. Oedipus believed that he had choices. He decided to leave home. He decided to become a brigand. He decided to kill the king and take the queen for his own. Or was it all illusory? Could he have exerted any choice at all once the oracle had appeared? Could he have remained at home and changed his destiny? Here, in the classic Greek tragedy, is exposed the dilemma between choice as reality and choice as illusion.

The Greeks came to realize that every choice is unique, that once a choice is made, a path of consequences unfolds, and that there is no way to go back and see what would happen if one had chosen differently. They realized that there is no evidence of free will — only a "feeling" — and that choice may be but a fleeting illusion.2

The ancient problem of free will versus determinism cannot be solved here, but it is a basic problem in environmental economics. If man is to manage his society, he must be able to choose among alternative economic and natural environments. If man cannot make these choices, he is limited to studying whatever economic and environmental circumstance he happens to be in.

Implied in the above remarks is a division of man's total environment into two categories — that which is controllable and that which is not. It is of extreme importance to understand that this distinction is purely conceptual. In reality, any environment is a tightly connected system. Because of this tight connectedness or interdependence, every act of choice or every manipulation of the environment sets off chains of repercussions, often uncontrollable, through the entire environmental system. For example, who, in the wildest flight of fancy, could have predicted the consequences of an innovation so apparently innocuous as the automobile? Yet, the automobile has disrupted families, poisoned the air, changed moral standards, moved whole populations off the land into the cities, destroyed the wilderness, and impoverished many who feel compelled to purchase these expensive machines. The automobile is quite ordinary as a device for moving people from place to place, but few inventions of man have had such profound social and environmental effects. The *intended* consequences of the automobile are quite admirable. Its *unintended* effects are pervasive and often highly destructive. *Environmental economics is the study of the unintended consequences of choice.*[3]

In sum, the environment of the human organism is a complex system of physical, biological, and social mechanisms that must continually adapt to the consequences of man's choices. While man is unique in that he can significantly determine his environment, he is similar to other organisms in that his behavior at any point in time is highly constrained by the environment he has created. Choice not only determines man's immediate welfare, it also determines the various options open to him in the future. For this reason it is essential that man choose well, that he take into account not only the immediate, intended consequences of his choices, but the long-range, unintended consequences as well. He must ask himself where he is going and if existing technology, values, and institutions will lead him there.

The evidence is not all in, but it appears that man is not always successful in getting where he wants to go. Indeed, the histories of developed nations of the world seem to be dramas with all the elements of a classic Greek tragedy. Act I is the long struggle between man and nature over the terms of existence. Act II, a consequence of the Industrial Revolution of the nineteenth century, introduces the essential element of hope. For the first time in human history, it seems possible that man will be able to lift himself above the ravages of poverty. In Act III, however, the classic resolution appears: Perhaps it is man's strength, not his weakness, his success, and not his defeat at the hands of an irresistible enemy, that poses the ultimate threat. In these initial scenes of the final act, it is becoming clear that economic success may result in social, environmental, and ecological collapse.

This book is a modest effort to call attention to the important choices facing man as continued economic success imposes an ever more serious burden on the natural environment. The choices are seldom clear-cut, so all that is offered is a frame of reference. A severe warning is in order lest the reader expect too much. The warning is made by describing what has not been attempted.

First, and certainly most important, no attempt has been made to provide a complete treatment of either economics or environmental problems. Instead, emphasis is placed upon establishing interconnections, and bridges between the two. There is the danger, of course, that the bridges will hang between two rather ill-defined and insubstantial points. But this route has been chosen in the interests of brevity and because there are many competent books in the two fields.[4] If such a comprehensive treatment were to be attempted, this book's small contribution would be lost in a morass of detail.

Second, this will not be a highly rigorous treatment of the subject. At this point in the ecological crisis, there is a need for a book that anyone can read. Also, there is no convincing evidence that either economics or the environmental sciences has evolved to the point where extreme rigor would help establish a bridge between the two. For the moment, the need is for more ideas rather than more logic. Both are necessary in the end, but the former must precede the latter. This does not mean that the book's ideas cannot be defended logically and empirically. It is only to say that not all that could be said is said here. This book is a primer, and like all primers, it needs additional and more rigorous development.

Finally, only a few analytical tools are presented. They are necessary for an understanding of the problems discussed. It is hoped that both the tools and the message are written in such a way as to be easily comprehended — even by those with no formal training in economics.

The theme of the book is that the current environmental problem is not taken seriously enough. This may sound strange in light of the plethora of doomsday predictions that appear continually in the news media. No quarrel is made here with the doomsday predictions, but perhaps "doom," in the environmental sense, should be redefined. Doom does not necessarily mean that man will soon cease to exist, but that he is perceptibly evolving toward the life style of the great majority of organisms on this planet — a life without variety or choice or beauty. This kind of doom may be much harder to avoid than extinction. Society will probably succeed in removing the conventional pollutants from the environment, and man will survive as an organism. But antipollution devices will not prevent the more insidious organismic doom that awaits us. Those who prophesy extinction are not sufficiently serious about the environment. The problems that must be overcome *merely to survive* are comparatively trivial, for survival can be achieved through manipulation of the physical world. But the environment is a social as well as a physical phenomenon. Indeed, alterations in the entire range of man's physical, biological, institutional, and social fabric may be required before the serious problems of the quality of life can be solved.

What must be faced is the kind of life toward which humanity is evolving and the enormous costs that must be incurred to change the direction, or at least the speed, of this evolution. The decisions made regarding the growth rate of the United States (or world) economy will determine the kind of life those living in the next century will have. If the present rapid rates of economic growth are continued, a "playboy" world of electronic gadgets and plastic habitat is virtually certain.[5] The natural world will be reduced to a kind of curiosum to be visited occasionally, much as people now visit "wildlife" in the zoo. If the decision is made to curb economic growth, however, a semblance of life as it is now known can be maintained. If growth is curbed, the idea that every man can be a "playboy" must necessarily be abandoned. Therein lies the basic problem: The deterioration of the environment is not a by-product of economic growth as so many believe; rather, it is in a fundamental sense a *direct product*. Economic growth is, in most instances, the conversion of the natural state of the world into a man-made state. Society is rapidly approaching the point (if indeed that point has not already been passed) where this free ride from the natural world is over. The major objective of this book is to elaborate on this theme — the connection between economic growth and the environment.[6]

Those who regard the environmental crisis as merely a matter of setting technology to work on air and water pollution, have misjudged the scope of the crisis. The seriousness of the environmental problem has also been underestimated by those who think it to be only a question of values and moral responsibility. Changed values and a new sense of moral responsibility are necessary ingredients of any solution to modern man's problems; but they are not sufficient. A nation of individuals with the best intentions can so structure itself as to make doomsday — defined even in that nation's own terms — inevitable. The awesome possibility exists that even while following prescribed paths to increased general well-being, the citizens of advanced nations may be severely limiting the prospects of the persons they most wish to help.

Although this is a pessimistic book, it is not fatalistic. Humans differ from other organisms largely in the range of choice they exercise. They lose the ability to choose and invite fatalism only when they believe they have no choice.

Because one never gets something for nothing, difficult choices must always be made. If an adequate level of environmental quality is to be restored and maintained, some conventional economic commodities will have to be given up. If the freedom to choose the kind of environment in which people want to live is to be preserved, the freedom of choosing to destroy the environment must be curbed. In sum, every choice brings benefits only at the expense of certain (often unanticipated) costs. It takes a great deal of thought and foresight to know whether as a whole and in the long run the benefits of any particular choice outweigh its costs. Determining the benefits and costs of choices is a central problem for those who wish to study the relationship between economic growth and environmental decay.

This book falls basically into three parts: The first part concerns the interconnections between economic growth and the environment. Chapter 2 discusses the history of growth and economists' skepticism regarding it. Chapter 3 examines growth in terms of resource depletion and the accumulation of pollutants in the environment. Chapter 4 inquires into the meaning of economic growth. It shows that the way in which growth is measured can lead to serious mistakes and also attempts to provide a more rational framework for an analysis of economic growth.

The second part of the book (Chapters 5 through 10) contains the central economic statement. Chapter 5 is a discussion of some philosophic difficulties economists have had to face when considering questions of values and tastes. Chapters 6 and 7 provide a set of economic tools and concepts that will be needed in understanding the subsequent chapters. Chapter 8 confronts the basic problem of "externalities" — those unintended consequences of choice that destroy the vital feedback mechanisms of economic systems. This theme is continued in Chapter 9 which shows how the economy and political systems are biased toward the production of private goods at the expense of the collective goods that constitute a large part of the environment. Finally, Chapter 10 argues that conservation makes good economic sense.

The final section moves into two arenas in which the battles of the war for environmental quality are likely to be fought, but in which economists have been very hard pressed to provide innovative approaches. In Chapter 11, nonmarket (or to some, "noneconomic") problems are discussed. This chapter builds heavily on the economic concepts of Chapter 7 and extends these arguments into new areas. Chapter 12 is more "activist." It deals with strategies and tactics in the environmental battle. Finally, Chapter 13 is a summary and integration. It returns to the basic problem of economic growth and environmental policy. It reasserts the view that high rates of growth are incompatible with high quality in the social and physical environment. It is possible to have one or the other, but, after a certain level of economic activity is reached, not both. The rest of the chapter discusses how the growth rate can be controlled, how problems of poverty and unemployment might be met despite low growth rates, and why, in an already prosperous nation, a stationary state of low or even zero economic growth is not only ultimately necessary but also desirable on its own account.

FOOTNOTES

1The effect of this ancient dilemma on philosophic and social development of the western world is explained by Popper in Karl R. Popper, *The Open Society and Its Enemies* (London: George Routledge & Sons, Ltd., 1945).

2The relationships between choice-making and economic theory have recently been elaborated by Buchanan in James M. Buchanan, *Cost and Choice: An Inquiry in Economic Theory* (Chicago: Markham Publishing Company, 1969). This is not a book for neophytes, but it clearly demonstrates the weakness of some prime linkages in economic theory.

3This definition is based upon Popper's classic description of the "characteristic problems of the social sciences" in Karl R. Popper, "Towards a Rational Theory of Tradition," *Rationalist Annual* (1949), pp. 41 and 42; "In all social situations we have acting individuals who do things; they want things; they have certain aims. In so far as they act in the way in which they want to act and realize the aims that they intend to realize, no problem arises for the social sciences (except the problem whether their wants and aims can perhaps be socially explained, for example by certain traditions). The characteristic problem of the social sciences arises only out of our wish to know the *unwanted* consequences which may arise if we do certain things. We wish to foresee not only the direct consequences but these unwanted consequences."

4There are many basic books in economics and ecology. We recommend Paul A. Samuelson, *Economics,* 8th ed. (New York: McGraw-Hill Book Company, Inc., 1970) and Eugene F. Odum, *Ecology* ("Modern Biology Series"; New York: Holt, Rinehart and Winston, 1963). By far the best source for information and background on the current environmental crises is Paul R. Ehrlich and Anne H. Ehrlich, *Population, Resources, Environment: Issues in Human Ecology* (San Francisco: W. H. Freeman & Co., 1970). One of the best collection of readings in this area is Glen A. Love and Rhoda M. Love, *Ecological Crisis: Readings for Survival* (New York: Harcourt Brace Jovanovich, Inc., 1970).

5The term "playboy" is intended to convey that obsessive need for the use and accumulation of "playthings."

6Because of the wide range of definitions of economic growth, the issue will be settled quickly, albeit arbitrarily. *Unless otherwise specified,* growth will mean an increase in per capita gross national product as defined in the national income accounting system. To be sure, this definition has some weaknesses. Many of the weaknesses will be discussed in Chapter 4. But for now, this common definition will be accepted.

chapter two

Economic Growth
And The Environment

It is difficult for a society to know where it is going until it knows where it has been. The purpose of this chapter is to provide a historical perspective on the relationship between economic growth and the environment in past civilizations as well as in the recent history of the United States, and to briefly examine the theory of economic growth developed by the Classical Economists. Hopefully, this historical sketch will provide an antidote to the belief that modern technology will provide an "invention" to "solve" whatever environmental problems may occur.

All growing economies face the problem of a deterioration in environmental quality. Often, a particular form of economic organization is blamed for permitting the deterioration to occur. In the United States, for example, the individualistic and highly dispersed incentive system is often blamed for the demise of the passenger pigeon, the cutting of the North Woods, the polluting of Lake Erie, and the smog that now hangs over the cities. These examples of decay are not problems peculiar to American capitalism but direct results of economic growth. Copenhagen has smog. Tokyo has smog. Leningrad has smog. Many beautiful relics of ancient Egypt are now under water as a result of massive dams, themselves symbols of economic growth. And all these circumstances have arisen because the societies involved — despite vastly different forms of social and economic organization — have been growing.

Any society must eventually realize that growth does not come free and that one of the first prices to be paid is some encroachment on the natural environment. Before New England could grow, the land had to be cleared, forests burned, habitats of wildlife ruined. Sadly, the faster New England grew, the faster the natural environment had to be destroyed. All nations, developed and undeveloped, are like New England. They face a very real choice between economic growth and environmental quality. A nation can grow rapidly if it is prepared to sacrifice its environment; *if the environment is to be maintained, lower rates of growth must be accepted*.[1]

Contrary to popular belief, economists have historically been skeptical of the long-run advantages of economic growth. Their underlying suspicions on the matter have constituted one of the few consistent themes in economic literature. True, in recent decades a few economists (along with Junior Chambers of Commerce, real estate speculators, and other civic boosters) have enthusiastically embraced "growthmanship." But most have remained doubtful about what has appeared to be growth-for-growth's sake.

With the publication of Adam Smith's *Wealth of Nations* nearly two centuries ago, economics became a systematic body of thought. Political economy quickly became a discipline that attracted such great scholars as David Ricardo, James and John Stuart Mill, Thomas Malthus, and J.B. Say. The Classical Economists (or Classicals), as they have come to be known, did not constitute a cohesive group. They had diverse concerns. Yet, in their collected writings, they anticipated nearly every modern economic problem.

On one theme in particular — economic growth — the Classicals showed an amazing consistency. They differed significantly on the *causes* of growth but only slightly on the *effects* of growth. Their theoretical system (or model) of growth rested on the notion that a specific relationship exists between economic progress and the ability of an economy to maintain those factors and institutions which contribute to that progress. In other words, their concern

12

was the ability of an economy to sustain a rapid rate of growth. Thomas Carlyle was so discouraged by the Classical model of growth that he described economics as "the dismal science." Placed in its briefest possible form and having all its frills and flourishes removed, the classical model of economic growth is as follows:[2]

1. A society begins with a natural capacity to produce an increasing amount of goods and support a growing population. That is, the society starts with an excess "carrying capacity" between its base of natural resources (mainly agricultural land) and its population. Increased use of natural resources creates a surplus of food, releasing laborers from food production and allowing them to turn to the production of tools (capital equipment). Tools further increase the productivity of labor, causing profits and savings to rise. Additional savings permits additional investment in even more capital equipment. And the process of economic growth continues.[3]

2. As a result of an abundance of food and other items necessary for life, population increases.

3. The only real obstacle to population growth is an emerging scarcity of food and other resources. Because of man's sexual instincts, population tends to grow to the limits imposed by the life-support system.

4. Ultimately, increases in population put pressures on economic growth. As the carrying capacity of the resource base is reached (that is, as resources become scarce relative to population), economic growth slows. Yet, population continues to grow.

5. As the imbalance between population and economic growth becomes more severe, savings fall, causing investment dollars to be used for food consumption. Profits and the economic growth rate tend toward zero. Standards of living fall back to subsistence levels.

6. Finally, zero population and economic growth prevails, perhaps at a higher level of total economic activity, but with no real improvement in the condition of the people.[4]

Thus, in the classical view, an economy proceeds through a full cycle, from subsistence levels through growth and back to subsistence levels. The gains from growth are only temporary; a period of "stagnation" inevitably follows the growth process.

There was ample historical evidence for the Classical theory of growth and stagnation. Though the pace was much slower and was often interrupted by war and pestilence, various ancient and medieval societies damaged or destroyed their environments as their populations and economies grew. The Greeks, for example, were wanton in their use of forest land. Nearly the whole area now making up Greece was once forested, but had been denuded as early as the Hellenistic period. (This is in contrast with Germans and other Central Europeans who developed an almost over-zealous worship of forests and who, in some instances, may have retarded their rate of economic advance because of their unwillingness to cut trees.) Although the explanation is highly speculative, the collapse of the Mayan civilization in Central America may have been a result of soil erosion. Much of the soil in that area is gone today, and one plausible explanation of the Mayan's demise is that population pressure may have required land to be cultivated too intensively each year. This overworking of the land eventually led to its deterioration in quality and the collapse of the social and economic system. On the other hand, the Incas of South America appeared to be more sensible. They fertilized their agricultural land with guano deposits mined on the Pacific Coast. This resource was so important to the survival of the Incas that it became a capital offense to kill the seabirds which provided this substance.

There is ample evidence of the pressure that economic growth places on a natural environment. For example, in China and England growth resulted in population pressure on agricultural land, in Greece, pressure on agricultural land as well as forests, in Western Europe, scarcity of agricultural land, in Spain, overgrazing denuded verdant areas leaving desert, in North Africa, wind erosion caused by removing native cover to extend agricultural development,[5] and in Ethiopia, excessive demands on forests for wood to be used in home building. The point is clear. In very early periods, technology was so primitive that man lived from hand to mouth. The economic progress of a society was intimately bound up with the ability of the society to produce food. In order to grow, an economy had to place increasing demands on the society's natural endowments.

Many people hold that, while centuries of recorded history may have been persuasive to the Classicals, modern economies are more or less exempt from the fate of older civilizations. Earlier societies were restricted by the resources at their immediate disposal. If their economic and social life was dependent upon forests, their civilization could flourish only as long as the forests lasted. If they were an agrarian people, progress lasted only until the agricultural land was exhausted (either in quantity or quality). But, it is argued, the rapid and widespread technological advances in the twentieth century have changed all that. Plastics have been substituted for wood and metal. Water and fertilizer have greatly increased the productivity of agricultural land and so have substituted for land in the production of food. Fossil fuels have been substituted for falling water and wind as sources of industrial power. The economies of the world have been "liberated" as these substitutes have become available. The optimistic view, then, asserts that an economy is unconstrained as long as science and technology grow as fast or faster than population.

Certainly, proponents of this view can marshal impressive evidence in support of their argument.[6] World population has expanded enormously over the past 200 years; yet, it is doubtful that one could point to any society which is poorer now than it was in the eighteenth century. And in a great many societies — in most of the world, in fact — people are substantially better off than they were before the Industrial Revolution.

Simon Kuznets has written about the relationship between population growth and economic output.[7] For his purpose, he dates the modern industrial period from 1750 and estimates a world population of 728 million at the beginning of the era (Table 1). This starting date was likely dictated as much by the availability of data as by anything else, but it does approximate the start of the Industrial Revolution. As the table indicates, the rate of population growth for the world has increased in each time period, 1750 - 1960. This is attributed to improved health and increased skills in providing the means of subsistence. On the North American continent, population growth increased demonstrably from the beginning, but, contrary to world trends, the rate of increase in North America has dropped since the earlier periods (with one exception).

Unfortunately, a table comparable to Table 1 cannot be developed for the history of world economic growth. But Kuznets has made some interesting population and economic growth estimates for the United States. In the period beginning in 1839 and ending in 1960 - 1962, population in the United States grew at a rate of 21.6 percent per decade. The gross national product (the total value of all goods and services produced in a nation in one year) grew much faster. During the period from 1839 to 1962, GNP grew at an average rate of 42.5 percent per decade.[8]

It is also interesting to note that the substantial rise in per capita GNP was accompanied by significant increases in labor productivity and reductions in man hours worked per capita. During the period 1850 - 1952, there was a 39.5 percent decline in length of the work year and a 24.2 percent decline in man hours worked per capita. In sum, the United States grew quite rapidly in population but nearly twice as rapidly in per capita output.

It is impossible to overstate the enormous increase in human welfare lying behind these statistics. The economic progress of the modern world is likely to be taken for granted today, but one needs only to read the history of a mere 100 years ago to appreciate the dramatic change in living conditions and working conditions brought about by economic progress. Describing living conditions of just 85 years ago, S. C. Burchell writes:

> In Belgium in 1885, the annual report of an inspector general of prisons and charitable institutions disclosed that the amount spent on feeding and clothing one convict exceeded the yearly wages of a whole family of workers — father, mother, and children. In 1860, a county magistrate in Britain reported that children employed in the Nottingham lace trade were "dragged from their squalid beds at two, three or four o'clock in the morning and compelled to work for bare subsistence until 10, 11 or 12 at night, their limbs wearing away and their humanity absolutely sinking into a stone-like torpor, utterly horrible to contemplate.[9]

The people of past decades and centuries paid an enormous price to raise themselves and their descendents out of poverty and to provide the basis for the affluence of the modern world. Few need to question the value of the economic growth which has brought the American and many other economic societies out of poverty. But the costs and benefits of *further* economic growth in a *prosperous* age are topics that should be seriously examined.

Much more must be said later about growth, population, and life-support systems. But before exploring these matters, other schools of economic thought, skeptical about either the prospects or the process of economic growth, will be mentioned.

One of these schools has been concerned with the stability of a growing economic system. Karl Marx (1818 - 1883) and Joseph Schumpeter (1883 - 1950) believed that the process of economic growth sowed the seeds of its own destruction, at least in capitalistic economies. Rather than a final state of no growth, they foresaw a complete breakdown of the system under the pressure of growth.

The great economist John Maynard Keynes (1883 - 1946) analyzed another distressing possibility. He showed how an economic system could reach a stable and continuing state of underemployment, a kind of permanent depression. Keynes' important discoveries inspired many of his contemporaries to develop their own "dismal" theories of growth. Alvin Hansen (1887 -) in expanding the notions of Keynes, reconstituted the classical theory of stagnation in terms of oversupply. He predicted the day when the consuming sector of the economy would become satiated with goods. Consumers would be unable to create sufficient demand to maintain any growth in the economy, and the economic system would eventually reach a state which Hansen referred to as "secular (or long-run) stagnation."

It is an interesting commentary on the times that what Hansen regarded as a clearly undesirable circumstance in the 1940's is

TABLE 1
North American and World Population Growth, 1750 - 1960

	North America[a]		World	
Year	Persons	Rate of growth per decade	Persons	Rate of growth per decade
	(millions)	(percent)	(millions)	(percent)
1750	1		728	
1800	6	43.1	906	4.5
1850	26	34.1	1171	5.3
1900	81	25.5	1608	6.5
1930	135	18.6	2015	7.8
1950	167	14.0	2509	14.3
1960	200	19.8	3010	20.0

[a]Includes the United States, Canada, Alaska, St. Pierre, and Miquelon.

SOURCE: Simon Kuznets, *Modern Economic Growth: Rate, Structure and Spread* (New Haven: Yale University Press, 1966), p. 38.

now, thirty years later, looked upon by some as, at worst, a mixed blessing and by others as a positive benefit. But Hansen did not pursue his analysis far enough to examine the fundamental issue of whether an economic system can survive secular stagnation.

Another view of the growth problem is held by such contemporary economists as R. F. Harrod, Evsey Domar, and Paul Samuelson. These analysts, generally, subscribe to the basic Keynesian theories, but they concentrate on questions regarding the dynamics and stability of economic growth. There too, results are depressing. These economists indicate that an economy can grow only along a very narrow and unstable path. In other words, a growing economy can easily be displaced from the growth path by any one of a host of possible disturbances. Once an economy is off this path, the system will tend to oscillate wildly between the extremes of severe depression and rapid inflation. Here, too, should be mentioned the findings of E. H. Phillips which lend credence to the conclusion that there is a necessary trade-off between full employment and price stability. The closer an economy comes to the attainment of full employment, the higher will be its rate of inflation. Price stability may be attainable only at the cost of high unemployment.

Partly for these reasons and partly on other grounds, there is a small but growing group of economists who are beginning to question the value of economic growth. They simply ask: "Why grow at all?" This group, perhaps founded by John Kenneth Galbraith, has already received the support of another well-known economist, E. J. Mishan, and a number of less prominent scholars. The ideas of these economists will be discussed at some length in the following chapter and throughout the remainder of the book. Their basic message can be presented very simply: the more developed nations of the world have now reached a state where all reasonable and rational demands for economic goods have been or can be satisfied. As a result, the virtues of added economic growth may be an illusion because growth does not come free. In fact, the costs of added growth are climbing quite rapidly as the pressures against certain resources, and on the environment as a whole, increase. The developed countries may have reached a level at which the costs of additional growth in terms of labor and loss of environmental quality exceeds the benefits, measured either in terms of increased economic goods or increases in GNP. The faster these economies grow, the poorer they may become.

Thus, in recent years an integration of the economic and ecological considerations begun some 200 years ago can be seen. This is long overdue, for had this formulation started a generation ago, the questions of economic growth and environmental quality might not be so urgent and so depressing today. Economic growth and ecological mechanisms are part of the same general system. They are not separate in fact; they cannot be comfortably separated in theory. Some of the specific links between the economic and ecological systems are explained in the next chapter.

FOOTNOTES

[1]This statement is to be taken quite literally. If resources are committed to the production of commodities, fewer resources are necessarily committed to the environment. Since this proposition stems from the fundamental economic concept of opportunity cost (explained below, p. 35), no country can escape it.

[2]An excellent introduction to the Classical model and the various growth theories mentioned in this chapter may be found in William J. Baumol, *Economic Dynamics: An Introduction,* 3rd ed. (New York: The Macmillan Company, 1970). The book presents a mathematical treatment of the subject and is thus particularly valuable to those with some mathematics background. But it is well worth pursuing even by those with little proficiency in mathematics.

[3]Both "savings" and "investment" have special meanings in economics. Savings occur when a person chooses not to spend some part of his income. To an economist, investment is not merely the decision by a household to purchase shares in a blue-chip corporation. It is the actual building of machines, trucks, factories, and other items that add to the productive capacity of the society. Investment must come from the part of income that is not spent — that is, out of savings.

[4]It will be seen later that John Stuart Mill deviated considerably from this viewpoint in his writings (Chapter 13).

[5]A number of authorities attribute the overextension of agriculture in North Africa to the introduction of the camel as a work animal and means of transportation. Similarly, the opening of the Central Plains States by the railroads led to the tragic dust-bowl episodes of the 1930's.

[6]One of the most consistently hopeful writers is Colin Clark, who for many years has contended that scientific progress could solve problems almost without limit. For one example of his optimism see Colin Clark, "Population Growth and Living Standards," *International Labour Review* (August, 1953). In this article, the author argues that there are no good economic reasons for supporting birth control legislation.

[7]The figures on population growth, growth in output, and input of labor and capital come from Simon Kuznets, *Modern Economic Growth: Rate, Structure and Spread* (New Haven: Yale University Press, 1966). In citing his work, it should not be implied that he subscribes to the optimistic view outlined above.

[8]The intricacies of GNP will be more fully explained in Chapter 4.

[9]S. C. Burchell and the Editors of Time - Life Books, *Age of Progress* (New York: Time - Life Books, 1966), p. 74.

chapter three

The Quantity
Of Product–
The Quality Of Life

The United States now has a gross national product of one trillion dollars. Many economists in the United States expect GNP to grow at an annual rate of about four percent in the near future. This means that GNP will double approximately every 18 years. In 40 years GNP would be more than four trillion dollars. In 100 years it would be 32 trillion dollars, and the twenty-second century would be ushered in with a GNP of 64 trillion dollars. If this trend continues, in 100 years the annual *addition* to GNP would exceed the *total* GNP of today's economy.

These calculations must be presented in conditional terms because the sheer magnitudes mentioned raise deep questions. Can the physical resources of the United States or world sustain such growth rates? Even if it is physically possible to grow this fast and to this degree, is it desirable in terms of the kind of life such intense economic activity requires? Can it happen, and if it can, is it worth it? These are the questions of concern in this chapter. The first is a physical problem of resource depletion and population expansion. The second concerns the quality of life and pollution, but in a larger and more complex sense than these terms are ordinarily considered.

RESOURCE DEPLETION

Economists of this century discarded the "dismal science" aspects of their heritage but evolved immediately into what might be described as an overly hopeful state. The Keynesian revolution seemed to solve the central problem of devastating economic fluctuations. Modern growth theory seemed to provide means of continuous economic progress. Skeptics bringing up the problems of population growth and resource depletion were almost universally considered holdovers from the dismal Classicals who had, through some fault of logic, missed the whole impact of modern developments, both in reality and in economic theory. The moderns, then, lived in a world of sophistication in economics, having as its purpose the final refutation of the existence of resource depletion.

The refutation was simple and direct. Economic growth could exceed population growth. Any number of people could be made progressively better off. There was no limit to economic growth because (1) technology would provide substitutes for whatever resources became scarce, and (2) as particular resources became scarce, their relative prices would rise, thereby discouraging their use and providing strong incentives to find and use substitutes.

Recently, however, this optimistic attitude has come under attack. The attack has been led not by economists, whose faith in technology remains virtually intact, but by the technologists themselves — specifically by the scientists with the responsibility for devising the resource-saving technology. Prompted by the skepticism of those in whom they believed most, economists are now beginning to reexamine factor substitution and technological advance. The results are creating some confusion among professional economists and at least a tentative return to the beginnings out of which the discipline grew. For the first time in 100 years, economic growth is being questioned.

M. K. Hubbard, the geologist, effectively described an essential point by suggesting that GNP is made from energy and matter resources. GNP is, in good part, a *physical* phenomenon. Therefore, it must be constrained by physical laws. These physical

laws simply do not permit constant exponential growth of anything — even GNP.

> One speaks of the rate of growth of GNP. I haven't the faintest
> idea what this means when I try to translate it into coal, and oil,
> and iron, and the other physical quantities which are required to
> run an industry. So far as I have been able to find out, the
> quantity GNP is a monetary bookkeeping entity. It obeys the laws
> of money. It can be expanded or diminished, created or de-
> stroyed, but it does not obey the laws of physics.[1]

The perspective provided by the laws of physics and of conservation is powerful. Clearly, when an economic commodity is
produced, it is not a pure act of creation as is sometimes implied.
The economic commodity is produced from energy and matter resources that may be exhaustible. Similarly, when an economic
commodity is consumed, it is not destroyed. It does not simply
vanish. It reenters the system either as materials to be recycled in
the production process or, as happens too frequently, as waste
materials detrimental to the environment. This integrated view of
the economic processes including production, consumption, and
reentry into a limited environmental system is causing reorientation of thinking among economists and others.

Much of the reorientation is beginning with a consideration
of the production possibilities open to an ever more crowded
planet. While it might be desirable to bring the production efforts
of all nations to levels currently enjoyed by the United States and
a handful of other highly industrialized nations, the resources of
the globe — *given current levels of technology* — will not permit
this to take place. The National Academy of Sciences has recently
published a very cautious report on the world resource situation.[2]
The report documents impending scarcity of nonfuel minerals
(mercury, tin, tungsten, and helium, for example) and warns that
new sources or substitutes must be found even if short-term needs
are to be met.

The report also notes that the world's known supply of recov-
erable petroleum liquids and natural gas will last only half a
century. The scarcity of natural gas brings into sharp focus the
problem of resource substitution. Current efforts to develop an
essentially smog-free automobile powered by natural gas may be
in vain if supplies of natural gas are as limited as the National
Academy report suggests.

Similarly, the effective lifetime of coal reserves as a principal source of energy is no more than two or three centuries. The Academy authors point out that "... we cannot simultaneously use the fossil fuels for fuels, petrochemicals, synthetic polymers, and bacterial conversion to food without going through them even more rapidly."[3] In addition to the global constraint, it must be recognized that the same resources cannot simultaneously be used for everything. Plastic houses save forests; bacterial food substitutes for traditional agriculture; natural gas substitutes for gasoline power. Each of these can be readily imagined; but all depend ultimately on the same fossil fuels, and it is possible that there may not be enough of these fuels to go around.

Equally depressing is the outlook for conventional sources of nuclear energy. The supply of Uranium 235 is low and low-cost production of nuclear power cannot be sustained for more than a few decades without dramatic breakthroughs in technology.

Continuing growth of GNP must be examined in connection with the kinds of resource scarcity mentioned above. It is important to realize that the enormous growth of the developed nations of the world has *not* been made possible through the advancement of technology alone. A supply of natural resources was also needed. Such developed nations as the United States, the European nations, and Japan either did not originally have or, in the course of their growth depleted, many of the natural resources that form the basis for their expanding industrial capability. Resource requirements of these industrial nations have been met by importing raw materials from less developed nations.

Some areas of the globe do not have sufficient internal resources to sustain a modern industrial economy even with the help of advanced technology. The existence of underdeveloped areas that can presently export raw materials and natural resources is pleasant and reassuring to the developed nations; but as these underdeveloped nations grow and require more of their own resources for their own industrial needs, their interest in exporting raw materials to developed nations is likely to diminish.

Estimates of the quantity of resources required for world industrialization are not easy to develop, but some speculation may be helpful. If the rest of the world were to attempt to attain the *present* United States standard of living (other things constant), it would have to increase its output from twice to fully fifteen times the United States' output. But this is only the beginning of the story. In developing its productive capacity, the United States has acquired a certain stock of capital in the form of machinery, buildings, highways, and the like. How much of this total capital would have to be developed in order to provide worldwide per capita GNP levels as high as in the United States is not known. Paul and Anne Erlich have developed some interesting statistics regarding the necessary raw materials. They estimate that raising the annual output of the rest of the world to United States levels would require 75 times as much iron, 100 times as much copper, 200 times as much lead, 75 times as much zinc, and 250 times as much tin as is now being consumed in world production.4

Moreover, the estimates of what would be required to bring the remainder of the world up to United States standards do not allow for continued economic growth. Population growth will put even more severe pressure on the world's stock of resources. Given the uncertainties attaching to each of these variables, it may be senseless to speculate further, for as Kenneth Boulding says:

> We do not really know the limiting factor. I think we can demon-strate, for instance, that in all probability the presently under-developed countries are not going to develop. There is not enough of enormous numbers of elements which are essential to the developed economy. If the whole world developed to American standards overnight, we would run out of everything in less than ten years.5

Finally, the strictly Malthusian specter has not lost its force. The food-producing resources of the world are not capable of feeding a continuously expanding population. Until the last dec-ade, many authorities anticipated massive famine in the 1970's. This date has now moved off into the indefinite future because of the astonishing, although perhaps temporary, success of the "green revolution."

> Between 1965 and 1969, land planted to the new varieties of wheat and rice in Asia expanded from 200 acres to 34 million acres, roughly one tenth of the region's total grain acreage. In one of the most spectacular advances in cereal production ever recorded by any country, Pakistan increased its wheat harvest nearly 60 percent between 1967 and 1969. This brought Pakistan, as recently as 1967 the second largest recipient of United States food aid, to the brink of cereal self-sufficiency. Progress for Pak-istan is not limited to wheat, for its record rice harvest in 1968

eliminated its deficit in rice, bringing it into the world market as a net exporter in 1969.

India's production of wheat, expanding much faster than the other cereals, climbed 50 percent between 1965 and 1969. Assuming political stability, present estimates indicate that India should be able to feed her vast and growing population from her own resources by 1972.

Ceylon's rice crop has increased 34 percent in the past two years. The Phillippines, with four consecutive record rice harvests, has ended a half century of dependence on rice imports, becoming a rice exporter. Among the other Asian countries that are beginning to benefit from the new seeds are Turkey, Burma, Malaysia, Indonesia, and Vietnam. Even such remotely situated Asian countries as Afghanistan, Nepal, and Laos are using the new seeds.[6]

The green revolution is not a permanent solution to the world's food problems. It is a means for buying time to get population under control, to develop even more advanced technology in food production and to change some attitudes about what constitutes a desirable way of life.

Like all technological change, the green revolution is likely to create severe problems of its own. First, it is creating and will continue to create huge supplies of cereals that will cause reductions in cereal prices as they reach the markets. These price reductions will lower farm incomes and, hence, the levels of living of some of the poorest people in the undeveloped countries. Second, new production techniques may also increase the use of pesticides and herbicides in world agriculture, adding to an already serious worldwide pesticide pollution problem. The technology of the green revolution must be considered a mixed blessing that provides short-term breathing space and a little time during which to attack the fundamental problems of population growth and income in the underdeveloped countries — a point emphasized by Norman Borlaug, 1970 recipient of the Nobel Peace Prize for his work on the green revolution. The National Academy of Sciences further emphasizes that

Foreseeable increases in food supplies over the long term, therefore, are not likely to exceed about nine times the amount now available. That approaches a limit that seems to place the earth's ultimate carrying capacity at about 30 billion people, *at a level of chronic, near starvation for the great majority* (and with massive immigration to the now less densely populated lands)!

... Hopeful allowance for ... [population] controls ... suggests that population *may* level off not far above ten billion people by about 2050 — and that is close to (if not above) the maximum that an *intensively managed* world might hope to support with some degree of comfort and individual choice. ... If, in fulfillment of their rising expectations, all people are to be more than merely adequately nourished, effort must be made to stabilize populations at a world total much lower than ten billion. Indeed, it is our judgment that a human population less than the present one would offer the best hope for comfortable living for our descendants, long duration for the species, and the preservation of environmental quality.[7]

Any discussion of resource depletion must end on a rather weak note. It is impossible to predict future resource use and future resource supply because it is impossible to predict future technological advances. At any given point in the future, resource scarcity may or may not be a major limitation to growth. But believing that technology will simply develop to solve whatever problems may occur is a blind expression of faith. No one can predict that technology will not appear, but neither can one predict that it will appear. In such a state of uncertainty and ignorance regarding the future, the rational approach is a conservative one.

THE QUALITY OF LIFE

Resources are not only depleted in physical quantities but also in quality. The deterioration of the environment so notable today is a reflection of the ecological system's inability to regenerate itself rapidly enough to stay in an acceptable state. Moreover, the system cannot absorb and recycle the continuing streams of waste products forced into it. The deleterious effects of waste products in the environment have received a great deal of attention. Often overlooked is the compound effect of chemical pollutants on biological phenomena. An example is the demise of Lake Erie in which complex interactions between chemical wastes and aquatic life rendered that body of water a dead lake.

Comparable events are continuously happening in the natural environment. Heat dissipated into a stream used to cool an industrial process may not harm existing aquatic life, but it may encourage the growth of algae or algae-like organisms which, in turn, may create havoc in the existing aquatic ecosystem. Since the threat of waste products to the health and continuation of our species has become a popular theme in the press and in public debate, only a few examples of these problems need be cited.[8]

One of the more striking cases is that of DDT and other non-degradable pesticides. DDT has come to permeate the air, land, and seas of the planet. Tests on the natural milk of a California mother revealed concentrations of DDT so high that this milk would have been barred from interstate commerce were it in any other container. Some of the effects of DDT on human and other organisms are well documented, and many are clearly harmful. It is known that DDT weakens the shells of bird eggs. Since many birds live at the top of food chains in the oceans, they are particularly susceptible to the accumulation of high concentrations of DDT in their systems. The weakened eggs break in the course of ordinary wear and tear, so the ability of these birds to reproduce themselves is seriously threatened.

Two other toxic wastes that pose serious health hazards are lead and mercury. Increasingly concentrated deposits of these materials are found in the air and water. Some indications of the more recent findings in this regard are outlined below.

Lead interferes with the basic metabolism of the body. It can seriously retard the operation of enzymes that aid in forming red corpuscles. Over 90 percent of the lead in the environment comes from automobiles burning leaded (ethyl) gasoline. Similarly, mercury in large doses can cause permanent brain damage or even death. The major sources of mercury are chemicals, plastics, paper pulp mills, and from its use as a fungicide in treatment of grain and other seeds. Mercury enters water through industrial wastes and, once in the water, is absorbed and concentrated in fish. Wild birds eat seeds treated with mercury and again the substance is concentrated in their systems. In 1969 the government of Alberta, Canada, closed the pheasant-hunting season because of excessive levels of mercury in these game birds. Similarly, fish caught in the streams and lakes of northern Canada have been found to contain high levels of mercury — endangering both the lives and economic activities of the Indian inhabitants of that area.

These are but a few instances of dangers associated with waste materials in the environment. The presence of these wastes detracts from the quality of life. It is easy to list other instances, but these are by now rather familiar problems, well discussed in other books, documents, and newspapers. In any case, the mere listing is not the major point. Even if all toxic elements were somehow removed from the environment and all the air was cleansed and the water purified, not all harmful impacts of economic activity upon the quality of life would be ended. Every act of economic importance produces good and bad effects. Each good or bad effect has physical and monetary aspects. A given automobile engine emits a known quantity of exhaust and thus produces a physical effect. The exhaust is harmful to those who must endure it, so it reduces the stock of values accumulating to them. Men place positive values on the engine, negative values on the exhaust.

As more automobiles are produced and used, the value of additional automobiles declines, but the cost (or negative value) of the exhaust these automobiles create increases. Moreover, there is apparently no automatic limit to this process. It is quite possible that an economy can grow to the point where additional costs exceed additional benefits. Net losses can accompany the process of growth.

One possibility for reducing the deleterious features of growth is to separate, through technology, the good features of growth from the bad. This "technical separability" can again be made clear by reference to the automobile. It is possible to design new smog control devices or new kinds of automobile engines (perhaps using steam or electricity) that achieve good effects (transportation) but with reduced bad effects (exhaust). The fact that there is a large number of possibilities to achieve technical separability in the automotive industry, in textiles, pulp mills, and other sectors of the economy provides one means for eventual control of much of the waste that pollutes the environment.

However, some important pollutants are not waste products or by-products; they are, rather, direct products for which the opportunity for technical separability is very modest. A perfectly noiseless, exhaustless automobile is still, in some contexts, a pollutant. A parked automobile, together with the necessary parking facilities, clutters the streets, destroys the scenery, and thus pollutes. Houses are very nice things to have, but they exist in space, thereby displacing space from some other use. Snowmobiles carry one through the winter landscape, scarring and damaging the terrain in the process. Factories, office buildings, power lines, highways, airports, and similar items are, *by virtue of their existence in certain contexts,* pollutants. Moreover, the separability characteristics of these pollutants — the possibility of keeping the goods and eliminating the bads — is very small. The goods and bads are and must always be produced simultaneously.

It may be objected that this is a naive, extreme view. First, it must be conceded that one cannot insist on absolutely untouched nature but must be willing to accommodate *some* change. It may also be granted that technological innovation of a different sort — of "urban planning," proper architectural "design," and related efforts — can substantially reduce the polluting effects of these production activities. Yet certain fundamental facts must be confronted: (1) there is a limited amount of space in a region, a country, or a planet; (2) the uses of this space are most often competitive; and (3) as the degree of economic activity grows, as the economy expands at rapid rates into the future, an increasing portion of the available space will be occupied by the products of economic activity.

The accumulation of people and their appurtenances in limited, technologically-nonexpandable space is perhaps the ultimate resource constraint and the ultimate problem of pollution. In the United States at one time there were vast areas of virtually free space. Growth proceeded with very little cost by expanding into this space. At the time it was not realized that this growth process was similar to the expansion of a culture in a sealed container, that eventually a point would be reached where the free ride would end. The vast tracts of space were progressively reduced to a few isolated pockets; then these areas were reduced to smaller numbers. Simultaneously, competition over the alternative uses of space became more intense: wilderness areas competed with mining and forestry; parks competed with office buildings; green belts competed with urban sprawl. Free space was the last frontier, and no technological innovation, whether of machine or design, can open it again.

In summary, it may be said that the problems of resource depletion and pollution are one: they are both problems of the conflicts of growth in a limited environment. Technological progress can alleviate some of the conflict situations. It can bring new techniques of production that require different resource inputs and alleviate some of the problems of waste products. It can even design the human habitat so that space is used more effectively. But as the incidence of the problem shifts from waste products, to providing consumer goods, to the congestion of people themselves, the range of technological possibilities narrows. The trade-offs between products and quality become more severe. It becomes increasingly difficult to have the cake and eat it, too.

FOOTNOTES

[1]Discussion by M. K. Hubbard in F. Fraser Darling and John P. Milton, eds., *Future Environments of North America* (Garden City, New York: The Natural History Press, 1966), p. 291. It is, however, important to realize that GNP is a reflection of values that *can,* in Hubbard's words, be "expanded or diminished, created or destroyed." The conservation laws impose constraints on the economic system; naturally they do not determine the whole of the system.

[2]National Academy of Sciences, National Research Council, Committee on Resources and Man, *Resources and Man: A Study and Recommendations* (San Francisco: W. H. Freeman & Co., 1969), pp. 5 and 6.

[3]*Ibid.,* p. 7.

[4]Paul and Anne Ehrlich, *Population, Resources, Environment: Issues in Human Ecology* (San Francisco: W. H. Freeman & Co., 1970), pp. 61 and 62.

[5]Kenneth E. Boulding, "Fun and Games with the Gross National Product," Harold W. Helfrich, Jr., ed., *The Environmental Crises* (New Haven and London: Yale University Press, 1970), p. 166.

[6]Lester R. Brown, *Seeds of Change: The Green Revolution and Development in the 1970's* (New York: Published for the Overseas Development Council by Praeger Publishers, 1970), pp. 4 and 5.

[7]National Academy of Sciences, *Op. cit.,* p. 5.

[8]These problems are surveyed in great detail in Paul and Anne Ehrlich, *Op. cit.,* Chaps. 6 and 7.

four*chapter four*

The Meaning Of
Economic Growth

In the preceding chapters, the focus has been on the social and environmental consequences of economic growth. Economic growth itself has been given only casual consideration. The definition and calculation of economic growth is not an idle matter. A nation may be seen to grow or not grow entirely as a result of what is included in its definition of growth. In the United States, growth has always been measured by changes in the *gross national product*; but GNP does not adequately take environmental considerations into account. This chapter details some of the shortcomings of the GNP method of measuring growth. It starts with a discussion of costs, builds through the national income accounting system, and ends with some suggestions for modification of current practices.

One of the fundamental concepts of economics is *opportunity cost*. When choices are made, one course of action is accepted while another is denied or rejected. The rejected course is the "cost" of the accepted choice. If steel is used for building bridges rather than for making ships, the nonproduction of ships represents the opportunity cost of the bridges. The ships are the benefit that has been foregone. The concept of opportunity cost, however, is not limited to things that can be measured in dollars and cents. For example, if a student must choose between studying and going to the movies, he experiences costs and benefits no matter which choice is made. If he decides to study, certain benefits are received and these must be weighed against the opportunity for pleasure given up by staying home. Similarly, if the choice is to go to the movies, the consequent benefit must be balanced against the costs of not having studied. Such decisions are difficult to make because a wrong decision is always possible — the costs may exceed the benefits.

The notion of opportunity cost applies to social as well as individual decisions. But when a society's choices are being studied, the analysis is more difficult. An individual can directly appraise costs and benefits in terms of their psychic impact on himself. Society, because its decisions affect so many people, needs a more objective (measurable) standard. One person cannot determine the value of movies and schooling to a whole society because he cannot compare different commodities in terms of their relative value to other people. If he is to be an intelligent participant in social affairs, however, he must attempt to estimate the value of commodities and services to other people through observation of their behavior or by some other means. Moreover, it is not sufficient to say that social well-being has benefited by the production of three automobiles, five movies, ten years of schooling, and two oranges and that this production process cost five tons of iron ore, two thousand hours of labor, and the destruction of ten trees and three birds. What is needed is a set of standards by which benefits and costs can be measured. Statements can then be made to the effect that benefits are equal to Y, costs to X, and that Y exceeds X by a certain amount. When the endless number of benefits and costs acquired and incurred by a complex economy are considered, the formidable nature of this accounting task can be appreciated.

Many good economists have worked for years to create and improve a system of accounts to aid in this important task. The result has been the national income accounting system of which gross national product is the most famous member. *The standard of value employed in computing the GNP is the price of goods and services bought and sold in the market place.* Thus, when it is reported that the GNP of the United States is approximately one trillion dollars per year, it means that the total selling price of all goods and services produced in the economy during a year is equal to one trillion dollars. Net national product, the other well-known item of the income accounts, is simply GNP minus the amount of capital used up in the course of the year in the production of GNP.[1]

Many difficulties have arisen in the construction and measurement of the national income accounts. Two central problems warrant comment. First, the accounts are limited to market-determined values since they include only the commodities bought and sold in the market place, thus, many nonmarket goods and services are not counted. Included in this nonmarket group are the labors of many housewives, home gardeners, and the value of the last few condors. Second, the national income accounts do not discriminate between costs and benefits, so they provide no way of determining *net* changes in social welfare.

The second point will be addressed first. Cigarette advertising increases GNP, since advertising is a service sold on the market. But it also encourages people to inhale poisonous substances — hardly a social benefit. Moreover, it does not seem to increase cigarette sales. Advertising provides a device to induce people to switch from Camels to Kools. Kools' gain is Camels' loss, more or less, but total industry sales are not increased. Yet, advertising expenditures by both companies are added to GNP. It can be argued that this is proper because advertising provides employment benefits for advertising employees and supports TV and much of the publishing industry. But would not society be better off if these resources were employed in other activities?

A similar phenomenon is the proliferation of gasoline service stations in the United States. Competition among the giant oil firms over their respective shares of this lucrative market has created a population explosion of gas stations. The consequence is that competition among the stations is so intense that each of them is grossly underutilized. Again, the construction of each new gas station increases GNP, but would not the economy be better off with a different allocation of resources?

Other examples may be mentioned. The pollution caused by economic activity does not *decrease* national income as measured by the accounts. In fact, if society is eventually forced to spend money to restore the environment, these expenditures also increase GNP. They do so in the same way that the production of food or medicine increases it. Or, finally, a tree left standing in a national park does not count in GNP; it is only when that tree is cut and sold as lumber that it is measured as an increase in GNP. The tree may be worth much more to society as a tree than as lumber, but that does not matter. It is "counted" only when it is sold as a marketable commodity.[2]

The idea that increased GNP (or NNP) is automatically a good thing is simply an illusion. Most people familiar with the construction of the national income accounts would not contend that these accounts provide an adequate measure of economic progress. The accounts do not estimate net changes in economic progress. They do not even separate costs from benefits. To attempt to maximize the growth rate of GNP would be akin to the frustrated businessman who decides that the ultimate goal of his enterprise is to maximize *expenditures.* He can devote all his energies to this end and obtain very large expenditures indeed, but he will become bankrupt in the process.

Enough has been said to indicate that rational decisions regarding economic growth cannot be made on the basis of the customary income accounts. These accounts do not effectively discriminate between what are properly considered costs of economic activity and what are properly considered benefits. It is beyond the purview of this chapter to attempt to discuss these problems in great detail. The remainder of this chapter, however, will attempt to provide a framework for new thinking on measuring the real growth (and real health) of an economy.

A society is properly interested in whether it is "richer" now than it was at some previous time. Assuming a constant set of objectives, the society is asking if these objectives are being satisfied better at time $t + 1$ than they were at time t. If this question is to be answered, the society must assess its position at t and at $t + 1$. It is then easy to determine the *net* change. But this process requires an estimate of changes in nonmarket positions as well as in market positions.

The income accounts recognize that in order to obtain net national product, the wear and tear on the capital stock that existed at the beginning of the period must be deducted from GNP. In principle this is correct. However, the only *capital stock* — the set of resources that can be used in more than one period — considered in this adjustment is *market* capital (machinery and equipment). Wear and tear on *nonmarket* capital should also be deducted.[3]

Nonmarket capital is the stock of "amenities" available at the beginning of a production period. A production period is begun with a certain stock and quality of air and water, open space, quiet, wildlife, natural beauty, health, and related items. Although these amenities are largely nonmarket goods, it is obvious that they contribute substantially to society's welfare. Indeed, the term amenity is unfortunate in that it implies a luxury; in fact, some amenities (air, water, health) are absolutely essential to human life. In any case, their value should be taken into account in an estimation of welfare. Amenity losses suffered because of the production of market goods should properly be entered as costs of that production.

Now what, exactly, results from a loss in amenity capital? The most obvious result is the expenditures made by persons attempting to restore their former levels of amenity enjoyment. For example, if one is accustomed to fishing and hiking as recreation and urban sprawl prevents local fishing and hiking, the costs incurred in reaching other fishing and hiking areas represent the cost of the loss of amenities. If, because of the effects of sprawl, the government must purchase additional national parks, this, too, is a cost of restoring some part of the amenity capital lost through growth. If air and water pollution create health problems, all expenditures on medical care due to this cause represent losses in the health amenities. If noise and smell force one into the suburbs or up to the fiftieth floor of a high-rise apartment house, the added costs of suburban living or high-rise living represent costs associated with amenity losses. The first approximation of the value of amenity losses may be taken to be the sum of all expenditures on substitutes for the amenities that were available at the beginning of the production period.

But cash outlays alone do not properly measure all amenity losses. Expenditures are simply substitutes. If one moves to the suburbs to buy clean air that has been lost in the city, the loss may not be fully reflected in the added expenditures he incurs. He may not enjoy the suburbs as much as he enjoyed being where he was; indeed, if he did, he would have lived in the suburbs in the first place. In other words, he will suffer a loss in addition to the increase in his direct expenditures. This effect is rather clear in the case of such "pollution diseases" as emphysema and cancer. Medical care for these illnesses costs a great deal, but the cash outlay is only a pale reflection of its real cost to the individual, his family, his friends, and society.

Another important characteristic of amenity capital is that it is most often a *stock* rather than a *flow*. When an economist speaks of a stock, he means an item that is used over more than one time period. A national park is a stock; the Ford Motor Company plant at Willow Run is a stock; and the United States Navy has a stock of destroyers. A flow is a resource or commodity that comes on to the market and is then taken off. Wheat and fish are flow commodities. The water in a flowing river is a flow resource since it passes on by. It must be used when it is present or not at all. Much amenity capital, because it consists of natural resources, is of the stock type. Professor Kenneth E. Boulding points out that there are many very puzzling questions regarding stocks and flows in economics. He is inclined to believe that the stock concept is more crucial:

> . . . we go to a concert in order to restore a psychic condition which might be called 'just having gone to a concert' which, once established, tends to depreciate. When it depreciates beyond a certain point, we go to another concert in order to restore it. If it depreciates rapidly, we go to a lot of concerts; if it depreciates slowly we go to few. On this view, similarly, we eat primarily to restore bodily homeostasis, that is, to maintain a condition of being well fed, and so on. On this view, there is nothing desirable in consumption at all. The less consumption we can maintain a given state with, the better off we are.[4]

Expressed in these terms, it is preferable to make as few incursions as possible into the stock of natural resources and amenities. The rapid increase in outdoor recreation expenditures in the United States may be interpreted as a measure of the depreciation of our natural capital stock of beauty, peace and quiet, and open space. The boom in the "control industries," from accounting and computers to police and fire protection (and related insurance costs), represents a similar loss in the initial stock of control and safety.

Without much difficulty, this line of reasoning can be extended. One million farmers will be forced off their land during the next decade. Many of them will have to find employment in the service industries. It would be interesting to assess the welfare losses experienced by these people and to compare these losses to the real contribution they will make to social welfare in their new roles in advertising, accounting, sales, clerical work, and the like. No one denies that the service occupations are necessary to the production of market products or even that some minimal level of these activities is absolutely necessary to any economic system. One does, however, question the net social gain represented by the fact that seven in every ten employees in the year 1980 will be in the service industries. As Leopold Kohr says:

> Once our life becomes too crowded and complex, a seemingly geometrically rising proportion of our output and consumption increase must be diverted from pleasurable enjoyment to the necessary but sterile task of helping us untangle the difficulties that have come to us as a result of our social, technical, and economic overdevelopment.[5]

Economic growth is partly a process of genuine productivity and partly a transformation of nonmarket goods and services into market goods and services. All of this requires painful adjustment. However, it would be hard to deny that over long time periods and up to a certain limit, human welfare has substantially improved as a result of economic growth as measured in conventional ways. Indeed, it is clear that the national income accounts seriously err on the side of underestimating many of the gains of economic growth, especially in underdeveloped countries. These accounts do not reflect the gains in health and longevity of the people, nor do they reflect the increase in leisure time made possible through rising productivity.

The basic source of error in the income accounts, both of over- and underestimation, is their failure to reflect the changing values of nonmarket goods. Food to a starving man is priceless; the amenities by comparison are almost worthless. If one has plenty of food to eat, the additional value of another steak is very low, while a shortage of amenities creates a very high value for these nonmarket goods. At this stage in the development of the United States economy, the income accounts tend to measure gains by counting increasing numbers of things that have lower and lower values. The accounts fail to include the mounting losses caused by reductions in amenities.[6] They provide a very distorted picture of where society stands and where it is going. Galbraith captured this distortion very well.

> The family which takes its mauve-and-cerise, air-conditioned, power-steered and power-braked automobile out for a tour passes through cities that are badly paved, made hideous by litter, blighted buildings, billboards and posts for wires that should long since have been put underground. They pass on into a countryside that has been rendered largely invisible by commercial art . . . They picnic on exquisitely packaged food from a portable icebox by a polluted stream and go on to spend the night at a park which is a menace to public health and morals. . . . Just before dozing off on an air mattress, beneath a nylon tent, amid the stench of decaying refuse, they may reflect vaguely on the curious unevenness of their blessings. Is this, indeed, the American genius?[7]

Mishan adds:

> Business economists have ever been glib in equating economic growth with an expansion of the range of choices facing the individual; they have failed to observe that as the carpet of "increased choice" is being unrolled before us by the foot, it is simultaneously being rolled up behind us by the yard. We are compelled willy-nilly to move into the future that commerce and technology fashions for us without appeal and without redress. In all that contributes in trivial ways to his ultimate satisfaction, the things at which modern business excels, new models of cars and transistors, prepared foodstuffs and plastic objet d'art, electric tooth-brushes and an increasing range of push-button gadgets, man has ample choice. In all that destroys his enjoyment of life, he has none. The environment about him can grow ugly, his ears assailed with impunity, and smoke and foul gases exhaled over his person. He may be in circumstances that he will never enjoy a night's rest at home without planes shrieking overhead. Whether he is indifferent to such circumstances, whether he bears them stoically, or whether he writhes in impotent fury, there is under the present dispensation practically nothing he can do about them.[8]

Neither Galbraith nor Mishan are directly addressing the problem of measuring national output, but together they do an excellent job of pointing out the absurdity of counting as our blessings only those things produced through the ingenuity of American technology and purchased in the market place. If the concept of economic growth is to be a viable concept, it must undergo substantial reconsideration. Although most of the nonmarket gains and losses resulting from economic growth cannot be readily counted or measured, they can be included in a *conceptual equation* such as the one developed below. The actual construction of such an index based on the equation is beyond the scope of this book. Hopefully, future research in the national income accounting area will result in a empirical approximation of the desired measure.

As a starting place for a conceptual equation, consider a society (or economy) that begins a production period with a given stock of capital equipment, *C,* and a given collection of amenities, *A.* The capital equipment consists of factories, dump trucks, dams, and all other man-made items used in the production of other goods and services. The amenities consist in all those attributes of the natural environment — qualities of water and air, wildlife, open space, and so on, discussed before.

Now in the process of producing GNP through the time period, a certain amount of the beginning stock of capital equipment and amenities are used up. As previously mentioned, that part of capital equipment destroyed in the process of production (CL) is *deducted* from the period's GNP to obtain net national product. This is reasonable because, if GNP had not been produced, this capital would not have been lost. So, in symbolic form

$$NNP = GNP - CL \qquad (1)$$

The task is to extend this basic equation into a more comprehensive measure; one that will also reflect changes in the nation's collection of amenities and similar nonmarket goods. The new measure will be called *net social welfare, NSW.* Several items that are not presently counted as contributing to the magnitude of GNP are positive contributors to the level of living or well being. Among these are some important nonmarket benefits of economic growth (such as added leisure time, the accumulation of knowledge, and improved health). These benefits will be designated *B.* Similarly there are costs associated with economic growth (designated growth costs, *GC*) that are presently included in measuring GNP but are not designated as costs. As an economy grows, more time, energy and dollars must be spent for information and control, highways, open space, and the reduction of pollution. At this time, all such expenditures are added to GNP even though they are expenditures which most often are made to restore a level of living rather than improve it. These costs should be deducted from NNP in order to more adequately measure movements in the well-being of a society. If these were the only changes, NSW could be designated as

$$NSW = GNP - CL + B - GC \qquad (2)$$

or

$$NSW = NNP + B - GC \qquad (3)$$

Amenities have not been considered. Growth, while incurring costs such as GC, may also cause the stock of amenities to deteriorate in much the same way that the production of GNP causes a reduction in the stock of capital. Reduction in amenities may occur as a result of increases in noise levels, increased smog, commercial development of scenic areas, and turning white water rivers into slack water reservoirs. Reflection will show that there is an intimate relationship between GC and the deterioration of the stock of amenities. If all noise is abated and if all smoke is effectively removed from the atmosphere, there will be no losses in amenities but GC will be very high. If no efforts are made at noise and smog reduction, GC will not be high but the stock of amenities of life will be substantially reduced. In sum, growth costs may not be high enough to maintain the quality of air, water, scenery, and other amenities through a production period. The result is that the quality of the environment may be lower at the end of the period than it was at the beginning. This remaining amenity loss, AL, should also be deducted from NNP if well-being is to be satisfactorily measured. This modifies equation (3) to

$$NSW = NNP + (B - GC) - AL \qquad (4)$$

and provides conceptual measure of net social welfare.

The relationships described by equations (1) and (4) can be examined in a hypothetical historical setting through the use of Figure 1. It is important to note that Figure 1 is drawn using only conjecture as a basis for both the relative shapes and the relative locations of the two curves *NNP* and *NSW*. Data are not available to allow more than this at present. At the time of the settlement of North America, very little was being produced by what was to become the United States economy. Additionally, life was rather harsh for the settlers so it is supposed that NSW was also very low. As the young economy grew, so did NNP, for more products were produced each year. With few exceptions, NNP has shown rather steady growth over the life of the nation and so is shown by the line *NNP*. Although the lefthand vertical axis does not show numeric values, *NNP* reached nearly $900 billion in 1970.

The time path of the hypothetical magnitude of net social welfare is shown as the curved line in Figure 1. It is highly speculative but it is likely that in the very early years, there were large nonmarket benefits *(B)* of growth and there were low growth costs *(GC)*. Because unsettled land was available and because increased growth resulted in more health, education, and (especially late in the nineteenth century) more leisure, it is not hard to imagine that over a very lengthy period of time, *NSW* grew faster than *NNP* — so much faster that it was undoubtedly higher than *NNP* for several decades.

Beyond point *y* (perhaps early in this century), the accelerating costs of economic growth in the form of amenity losses and growth costs began to reduce the difference between *NSW* and *NNP*. Both were increasing, but *NSW* began to increase at a lower rate than *NNP*. In other words, progressively more *NNP* had to be generated in order to obtain a given increase in *NSW*.

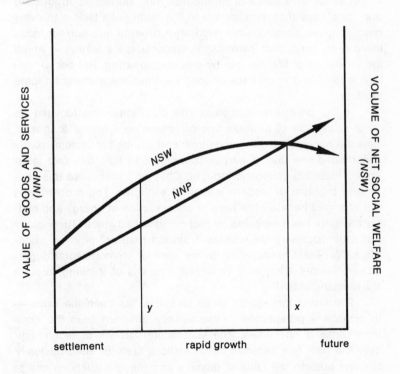

It follows that if these trends continue over a sufficiently long period of time (which may have already arrived), the growth in costs will equal the growth in benefits. At this point, *x*, *NSW* will no longer increase even though *NNP* continues to increase. The costs of growth equal the benefits. It is also possible that the economy could proceed even beyond *x* to a point where costs exceed benefits and where *NSW* declines even as *NNP* continues to increase. It may be that the U.S. economy is rapidly approaching point *x* — if, indeed, this point has not already been reached. The public concern for environmental quality, the increase in research on pollution and its control, and even the many local efforts to clean up small areas may be visible signs of a feeling that NSW is declining. Just prior to leaving the United States after a long career as a writer and humorist, S. J. Perelman perhaps best captured the present condition of society by describing New York and its environs as a "terminatory."[9] Perelman uses the word not in the sense of a place of termination but, rather, as suggesting the "ambience that termites live in." A humorist's task is to overreact to given situations but even after brought into correct focus there is no doubt that Perelman is discussing a situation in which the amenities of life are one by one disappearing. Net social welfare is declining in the face of continued increases in net national product.

In the past three chapters, the discussion has focused on various aspects of a single line of reasoning. Chapter 2 showed how the balance between economic and ecological systems partly determined the fate of earlier societies and how this fact influenced classical economic thought. Chapter 3 discussed this continuing problem in modern industrial societies. The problem has not changed because the laws of conservation of energy and matter have not been repealed by technological advance. Technology has only displaced the problem from the realm of physical apocalypse (at least temporarily) to the area of environmental degradation. Finally, Chapter 4 provided a means of thinking through the illusions of GNP.

The intent throughout these chapters has been the same — to provide a perspective fundamentally different from the common sense of the times. The spectacular growth of technology over the past few decades has created a state of mind basically counter both to the laws of physics and the life sciences and to the principles of economics. It is expressed in the idea that growth is free, that somehow goods are created from nothing and, after having yielded their benefits, vanish into nothingness once more. Modern growthmanship is a utopian fallacy resting on the premise of "something for nothing." The argument here has been that, on the contrary, for every benefit there is a cost; that in any growth process the value of additional benefits declines as the costs of obtaining them rise; that eventually a point is reached where the additional costs equal the additional benefits. That point is the place to stop. Perhaps modern industrial society has reached the stage where the direct benefits of economic growth are being overwhelmed by the indirect costs of environmental degradation. That is the economic lesson.

Ecologists arrive at the same basic conclusion. Social systems differ from biological systems in certain respects, but they resemble them in others. One similarity is the necessary equilibrium between internal structure, size of system, and external environment. For example, a human cannot be ten times his present size and maintain the same proportions. If every part of a man were increased ten times, the skeletal structure would collapse under the weight of the body. During the age of giantism, dinosaurs gradually evolved to enormous size; but, in attaining that size, they became very specialized organisms, unable to adapt to altered environmental conditions. The growth of social organisms obey the same basic laws, although the processes of growth and adaptation are, of course, different and much more complex. Humans can choose to grow economically, and they can choose to adapt to altered environments or even to alter their environment. They have free will. But the danger is that they will not make the necessary choices until it is too late. The benefits of growth are apparent; the costs of growth are subtle and insidious. As growth continues, the strain on the internal structure of society increases. In order to withstand that strain, society becomes rigid and specialized. It loses its flexibility and adaptability — what was a strength becomes a weakness. Today's society, because of a refusal to acknowledge the consequences of growth, is in considerable danger.

FOOTNOTES

[1]For further elaboration of the national income accounts, see Paul A. Samuelson, *Economics,* 8th ed. (New York: The McGraw-Hill Book Company, 1970). Chapter 10 deals with the income accounting system.

[2]See, also, A. A. Berle, Jr., "What GNP Doesn't Tell Us," *Saturday Review* (August 31, 1968).

[3]This approach is suggested in Ezra J. Mishan, *The Costs of Economic Growth* (New York: Frederick A. Praeger, 1967), Part 3.

[4]Kenneth E. Boulding, "Economics of the Coming Spaceship Earth," Reprinted in Love, *Op. cit.,* p. 314.

[5]Leopold Kohr, "Toward a New Measurement of Living Standards," *The American Journal of Economics and Sociology* (1955), p. 93.

[6]Here the discussion is anticipating a vitally important concept explained more fully in Chapter 6. This is the concept of the "margin." It is not the total value of all economic activity nor of all the amenities that is in question; it is, rather, the marginal value — the value of a little more or less. Thus, food can be "priceless," but once one is reasonably well fed, the value of another steak may be rather low.

[7]John Kenneth Galbraith, *The Affluent Society* (New York: The New American Library, 1958), pp. 199 - 200.

[8]Mishan, *Op. cit.,* pp. 85 - 86.

[9]Phillips McCandlish, "A Humorist's Bitter Adieu to the U.S.," *New York Times* (September 18, 1970), p. 45.

part two

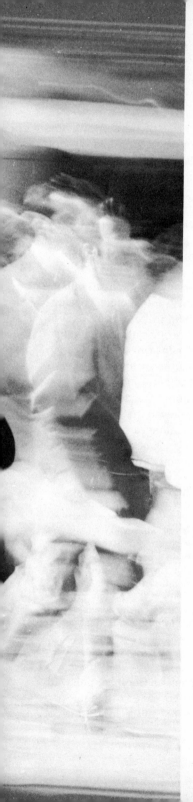

chapter five

Preferences And Values

THE PROBLEM

OF VALUE

When the Lord looked out on His work and found that it was good, He enjoyed a prerogative not given to mortal man. Mortals must always ask questions of value — good for what and good for whom? A great deal of effort has been expended in an attempt to find those attributes that give an object value and many philosophic ships have been wrecked on these rocks. Despite the intellectual difficulties, the problem of value is one that cannot be ignored. The central message of this book lies among the value conflicts that arise when a commodities-oriented economy develops the desire to restore and maintain its quality of life. This small book provides neither the space nor the opportunity to treat the deep philosophic aspects of this problem with any degree of thoroughness. Instead, in this chapter an attempt will be made to explain how economists think of value, to describe the pitfalls involved in this way of thinking, and to point out some promising paths that are beginning to open as philosophers and economists probe this problem.

Oscar Wilde described the cynic as "one who knows the price of everything and the value of nothing." With this, he not only put cynics in their place, but he also clearly defined the interesting conflict between the values men hold and the prices they pay. This conflict is particularly disturbing to economists. Over the years, they have become accustomed to thinking in terms that have substantially clouded any clear distinction between value and prices. To the displeasure of philosophers, economists speak almost indifferently of price theory as "value theory."[1] But this is getting somewhat ahead of the message.

A useful place to start is with Jeremy Bentham. Bentham (1748 - 1833) was a major figure in English intellectual development. Like all intellectual leaders, Bentham owed a great deal to what went before him and much of what is often attributed to him is hinted at in the writings of Adam Smith, David Hume, John Locke, and other English intellectual figures of that era. Nevertheless, Bentham pulled a number of rather vague ideas into a grand synthesis that has substantially altered the development of economic thought and policy. Bentham believed that man is driven by the forces of pain and pleasure. In Bentham's estimation, human behavior can only be understood in terms of man's conscientious avoidance of pain and avaricious pursuit of pleasure:

> Nature has placed mankind under the governance of two sovereign masters, pain and pleasure. It is for them alone to point out what we ought to do, as well as to determine what we shall do. . . . They govern us in all we do, in all we say, in all we think: every effort we can make to throw off our subjection, will serve but to demonstrate and confirm it. In words a man may pretend to abjure his empire: but in reality he will remain subject to it all the while. The principle of utility recognizes this subjection, and assumes it for the foundation of that system, the object of which is to rear the fabric of felicity by the hands of reason and of law. Systems which attempt to question it, deal in sounds instead of senses, in caprice instead of reason, in darkness instead of light.[2]

If man is a lusty and acquisitive beast, as Bentham believed, it is only because circumstances make him so. Sex, hunger, and the single-minded pursuit of money were rather low forms of pleasure in Bentham's estimation. They are not so much forms of pleasure as a means of escape from pain. Bentham, himself, looked forward to the day when the physical constraints of the world would be relaxed sufficiently to enable man to pursue the pleasures of intellect, arts, and the good life. Now, with the advantage of hindsight, the intriguing nuances of Bentham's expectation can be seen more clearly. For most people in Bentham's western world, survival is no longer a major problem, but the struggle for it has become a way of life. Man may not have to struggle as he once did, but he continues to, one suspects, because he knows no other mode of behavior.

Bentham had a strong practical bent and thought all things should be useful to man. Indeed, his philosophy was called *Utilitarianism*. Bentham, himself, carried his belief to somewhat eccentric extremes.[3] An anecdotal account relates that, while sitting in his study one day, a queer waste of resources occurred to him. On the one hand, he observed, there are sculptors laboring away, chipping at lumps of rock in an attempt to transform these rocks into statues of great men to be used as decorative and inspirational figures in public parks and before public buildings. On the other hand, lying in their graves, serving absolutely no good purpose, lay not the mere likenesses of great men but the actual originals! Why not mount these bodies and use them in the public parks and before great buildings in place of the more conventional statuary? Hordes of sculptors would be released for higher-valued uses (such as nonrepresentational art), and undervalued resources, hitherto occupying otherwise perfectly good land, would be utilized.

Bentham was no idle dreamer; he was a man of action. Being a great man in English letters, he dedicated himself to the cause. Thus, in accordance with his will, Jeremy Bentham was stuffed and mounted. With Bentham as with so many others, however, there has been a certain lag between the great idea and its common acceptance. Bentham does not stand in his glory before the houses of Parliament, nor even in Hyde Park. He has been ignominiously relegated to an obscure closet in the University of London. There he stands to this day, his head between his feet, contemplating a backward world.[4]

If Bentham were to come out of his closet and travel around the Western world, he likely would not be pleased with all he would see. But it is doubtful that he would recommend that very much be done about it in the way of public action. For Bentham's philosophy rested on two cardinal principles that effectively limit the role of the state in economic or social affairs. He believed that men should get what they want and that the individual is the best judge of what he wants. It seems to follow that, so long as one man's pursuit of happiness does not interfere with another man's pursuit of happiness, everything should be left alone. From these principles Bentham concluded that the best society is the one that provides the greatest good for the greatest number.[5]

Bentham's views on the structure of society made him the original "do your thing" man. Western economics following closely on his heels, evolved as a science that depends on man behaving rationally and in his own best interests. The most desirable economic system is the one that best serves the preferences of individual consumers. The market system fails when it does not serve these preferences. Whatever contributes to the satisfaction of individual preferences is decreed good, and whatever detracts from their satisfaction is bad. This principle has become known as the doctrine of "Consumers' Sovereignty." Thus, Bentham has become an intellectual, although not an ornamental, monument. It must give him some comfort in the confines of his dark closet to know that, after so many years of neglect, at least economics is securely cast in his image.

While Utilitarians developed the foundations of the theory of economic value, it was up to later economists to refine and extend it. One very important development in the theory of economic value stems from the work of Vilfredo Pareto, an Italian sociologist-economist-mathematician. Out of Pareto's original work evolved modern "welfare theory" — a set of economic concepts that attempt to explain in precise terms the necessary and sufficient conditions for an economic optimum.[6] According to this theory, an economy is at its optimum when it is impossible to make anyone better off without making someone else worse off. This statement sounds like a truism, but behind it lies a wealth of interesting and elaborate arguments. Of particular interest is a concept which may be best approached through example. Imagine that a new labor-saving machine is invented, causing laborers to lose $1000 in wages. But because the new machine is more efficient than the old hand labor method of production, profits to businessmen rise, and consumers can buy the product at lower cost. The increased profits plus the costs saved by consumers result in a total economic gain of $10,000. The question is: Does society as a whole gain by the use of this machine?

According to the Utilitarian criterion of the greatest good for the greatest number, it appears that society does gain. A great many people would benefit by ten times the amount lost by those few laborers who are now out of work. It can also be argued that the loss of $1000 by the laborers causes suffering and poverty that is not sufficiently offset by the gain of $10,000 to affluent businessmen and consumers. Unfortunately, this is not a testable proposition. It is a matter of opinion whether or not society is better off.

According to the "Pareto-optimality" criterion, it is definitely impossible to declare that society as a whole has been improved by the use of the machine. Some people are better off; other people are worse off. But the two groups cannot be compared. The machine fails to pass the test for an improvement in society as a whole. No sophisticated modern economist would be prepared to say that on this evidence alone such a machine should be used, even though there is $9000 increased income (or "savings") from its use. But this criterion leaves economics at something of an impasse, for almost any change in an economic system leaves some people better off and some people worse off than before. The criterion seems to lead to the conclusion that the status quo is always best.

In order to overcome this impasse, two distinguished economists, Nicholas Kaldor and John R. Hicks, developed what has come to be known as the "compensation test."[7] This test is one of the major concepts of modern economics. It says that a change from one state to another is desirable only if those who gain from the change can compensate those who lose. The compensation must be to the full extent of the losses, and it must leave the gainers better off than they were before.[8] If businessmen and consumers were to take part of their $10,000 gain and pay it to the laborers in compensation for their loss (which may exceed the $1000 loss in wages) *then and only then* would the machine be considered an improvement for society as a whole. The machine would pass the compensation test and the Pareto-optimality test. Some people would be better off (businessmen and consumers), and no one (laborers) would be worse off than before.

The Pareto-optimality criterion and the compensation test are powerful tools in any study of economic problems — especially those related to the physical environment. Using these concepts, for example, it will be possible to show how the pollution of the environment leads to economic inefficiency and how insistence on compensation can result in net gains to society as a whole. With excessive amounts of pollution, society is not at a Pareto-optimal state. "Everyone" can be made better off by reducing pollution, and one way to reduce it is by forcing polluters to pay compensation to those who suffer from it.

These concepts form the foundations of modern welfare economics and are, thus, integral to the study of environmental economics. However, both Utilitarian analysis and modern welfare theory share a basic problem that has never been adequately resolved — the doctrine of consumers' sovereignty. According to this doctrine, the ultimate source of value is consumers' tastes. These tastes distinguish good from bad in the economic and social systems. The doctrine implies that whatever people want is right. While almost everyone would agree to this proposition up to a point, very few would accept it without limit; yet, the limit has never been adequately defined. It has not been determined just how sovereign a consumer can or should be.

Critics have frequently attacked economics by saying that man is not a rational, calculating creature who knows what ends he wishes to serve. Rather, the critics say, man tends to be a creature of habit, driven by sometimes irrational forces. Whatever the merits of this argument, the critics have missed an essential point by failing to note that the ends themselves may be irrational or undesirable. This is perhaps the greatest weakness of contemporary economics. By taking ends as given and rational, the subject matter loses its grip on many of the more vital dimensions of human welfare. In his presidential address to the American Economics Association in 1968, Kenneth Boulding noted this shortcoming in economic analysis.

> One of the most peculiar illusions of economists is . . . that tastes are simply given, and that we cannot inquire into the process by which they are formed. This doctrine is literally "for the birds," whose tastes are largely created for them by their genetic structures, and can therefore be treated as a constant in the dynamics of bird societies. In human society, however, the genetic component of tastes is very small indeed. We start off with a liking for milk, warmth, and dryness and a dislike for being hungry, cold, and wet, and we do have certain latent drives which may guide the formation of later preferences in matters

of sex, occupation, or politics, but by far and away the largest part of human preferences are learned, again by means of a mutation-selection process. It was, incidently, Veblen's principal, and still largely unrecognized, contribution to formal economic theory, to point out that we cannot assume that tastes are given in any dynamic theory, in the sense that in dynamics we cannot afford to neglect the processes by which cultures are created and by which preferences are learned.[9]

As Boulding points out, Thorstein Veblen, writing early in the twentieth century, was virtually the only economist who attempted to penetrate beyond the assumption of given tastes into a dynamic theory of how tastes are formed. This theory led him to profound skepticism regarding the ethic of consumers' sovereignty.

Veblen's theory can be put very simply. Everyone desires the good opinion of others. This good opinion is largely a function of the amount of material things one can conspicuously consume. But the people whose good opinion one desires most are those whom one holds in most esteem — that is, those who conspicuously consume more than oneself. One is induced, therefore, to spend up to the level of one's peers. But this level, being always the next highest, is continually rising. Thus, one is driven to the expenditure of ever larger sums of money in order to augment one's status or, if the others are also expanding their expenditures, simply to preserve it. Veblen dared to wonder if this ever-expanding cycle of conspicuous consumption constituted the best of all possible worlds.[10]

If Veblen's notion that conspicuous consumption is the main cause of high standards of living is correct, a depressing conclusion immediately follows. Man will be driven to ever higher levels of consumption with no real increase in contentment. In the conventional view, consumption is necessary to meet certain material needs and acquire a level of comfort. It assumes that a limit can and will be reached. People can become, if not satiated, at least so comfortable that the need for further consumption is very low. Veblen's theory shows that this is not necessarily the case.

Conspicuous consumption results in the kind of perverse dynamics familiar in international power relations where each side must continue to expand its military might in order to negate the expansions of the other side. Since both sides expand in reaction to the other, the level of security enjoyed by each either remains unchanged or may even decline as each becomes more powerful. All that increases, in the end, is the level of expenditures necessary to maintain a given level of security.

So it is, according to Veblen, with modern habits of consumption. More is continually needed, not necessarily because of the real satisfaction it brings, but because without it one will have less than his neighbors. And that, under the canon of conspicuous consumption, is a fact too horrible to contemplate.

The dismal conclusions of the theory of conspicuous consumption naturally lead to questioning the assumption of "given ends." If people come to realize the futility of this perpetual round of expenditures, they may change their habits. After appraising their own welfare in more comprehensive terms, they could decide that their tastes are *wrong*. Such a decision would not only alter personal behavior but would require a reexamination of the fundamental principle of consumers' sovereignty, as well. Many modern economists are interested in undertaking such a reexamination but where to start is a perplexing matter.

The complete abandonment of the concept of consumers' sovereignty is surely too severe. Everyone should, within limits, "do his thing." What those limits are, exactly, is a very difficult matter to describe and would require clear-cut definitions. The liberal philosophy of the Western world holds that one should be able to do what he wishes, providing (1) that he does not harm others and (2) that he is not likely to harm himself irreparably. The former condition is stronger than the latter because, while it is generally easy to tell when one is harming others, it is very difficult to tell when one is harming himself. If an industrialist pollutes the water, it is clear that he should be restrained for others are being harmed. If an individual pollutes himself with nicotine, marijuana, or heroin, the role of the state is less clear.

In sum, questions of value depend upon choice. Without genuine choices among alternative courses of action, the problem of what one "ought to do" is meaningless. The more choices a person has, the more likely he is to be able to satisfy whatever values he is pursuing. Without a wide spectrum of choice, even the doctrine of consumers' sovereignty is meaningless. Why be sovereign if you cannot choose? It would seem that a necessary function of the good society is to provide for its constituents the maximum possible range of choice among goods, services, and situations.

The provision of this expanded range of choice sounds very much like an extension of the doctrine of consumers' sovereignty.

People will still choose on the basis of their tastes even in a society with greatly increased numbers of alternatives. But there is
a significant and fundamental difference. If society is to have as
one of its goals the maintenance of wide ranges of options, an
individual's behavior can be evaluated on the basis of what his
choices do to his choice-making capacity as well as what his
choices do to his immediate well-being. A useful example is the
profligate who spends himself into oppressive debt. According to
the doctrine of consumers' sovereignty, he is simply following his
tastes. The criterion of expanded choice shows, however, that he
is behaving in an irrational manner, for his actions are destroying
his range of choice (as well as his ability to act rationally in the
future). This is the criterion underlying many apparently sensible
social restraints on individual choice that would not otherwise
make sense. It is (now) illegal to sell oneself into indentured serv-
itude; that is, to make an irrevocable contract to work for someone for a certain number of years. It is illegal to attempt to com-
mit suicide. It is illegal to consume heroin or to fail to provide
minimal schooling for one's children. Society even provides a
way out for the profligate; he can legally become bankrupt and
be relieved of debts he cannot pay. In other words, it is possible
to do what one wishes so long as it does not harass others and
unduly curb one's own range of choice. The individual is free to
do what he wants so long as both he and everyone else remain
free.

While consumers' sovereignty is an important element of the
good society, it is not the only one. Consumers' sovereignty is
meaningless without choices, and individual tastes can affect the
range of choice available. Tastes can and must be appraised in
terms of freedom.

The criterion of expanded choice provides some insight into
contemporary "environmental" controversies (see Chapter 10).
One need not despise the very real advantages of the great cities
of this nation to enter a plea for the preservation of the natural
environment. Nor need one despise the city to protest the fact
that rural areas are being swept clean of their populations, and
that people are tending increasingly to live in cities, often against
their will. Instead, it is sufficient to insist that people should have
a choice in anything as important as where they live. Within reason, jobs should be provided where people want to live; people
should not be forced to live where jobs happen to be. Nor is it

absurd to believe that the environment should be preserved for the enjoyment of future generations. Perhaps it is true, as some allege, that future generations would adapt to an artificial world and never miss what they have never known. But one thing is certain: Unless present generations preserve this option for them, they will never have the chance to decide. They will not have the choice. In its essential features, the environmental movement is a plea for increased choice; it is a plea for freedom.

FOOTNOTES

[1]The early classical economists were very concerned with this problem. They carefully distinguished between "value in use" and "value in exchange." Later, with the development of new economic theories, value in use was largely ignored. The tradition extends, with few exceptions to the present day. There seems to be a feeling that while value in exchange is an economic problem, value in use is a problem outside the discipline's bounds. Hopefully, current work by philosophers as well as economists will help clarify the distinctions.

[2]Jeremy Bentham, *An Introduction to the Principles of Morals and Legislation* (Oxford: Basil Blackwell, 1948), p. 125.

[3]A complete account of the following is found in Nigel Dennis, "A Treasury of Eccentrics," *Life* (December 2, 1957), pp. 101ff.

[4]His head was amputated for anatomical dissection, a use Bentham would have heartily approved. Rumor has it that, upon inquiring into Bentham's skull, the anatomists found an abacus-like structure where beads moved up and down rods clearly marked with cardinal measures of pleasure and pain. But since these results have not been duplicated by independent investigations, they have been suppressed. John Von Neumann and Oskar Morgenstern in *Theory of Games and Economic Behavior* (Princeton: Princeton University Press, 1944), however, have reported the observation of cardinal utility. With this evidence, the task of the next generation of vivisectionists is clear.

[5]The operationally impossible term "greatest good for the greatest number" is often ascribed to Bentham and is often used as a capstone for his elegent Utilitarian philosophy. Even though the term inspired Bentham, it was apparently not his invention. It is more often attributed to an earlier scholar, Joseph Priestly, in his *Essay on the First Principles of Government* (London, 1768).

[6]The term welfare theory defines that branch of economics relating to the exposition and appreciation of economic well-being or welfare. It is a comprehensive body of theory and quite different from the layman's conception of welfare that is most often confined to public welfare programs.

[7]Nicolas Kaldor, "Welfare Propositions in Economics," *Economic Journal,* vol. XLIX (1939), pp. 549 - 52; and J. R. Hicks, "Foundations of Welfare Economics," *Economic Journal,* vol. XLIX (1939), pp. 696 - 712.

[8]Economists commonly use the expression "everyone is better off" to describe a change which leaves someone better off and no one worse off than before.

[9]Kenneth Boulding, "Economics as a Moral Science," *American Economic Review,* vol. 59 (March, 1969), pp. 1 - 2.

[10]Although Veblen's writing spanned a period of three decades, the essential arguments for present purposes are found in Thorstein Veblen, *The Theory of Business Enterprise* (New York: Charles Scribner's Sons, 1904), and *The Theory of the Leisure Class: An Economic Study of Institutions,* (New York: The Macmillan Company, 1899).

chapter six

Two Tools Of Economic Analysis – Supply And Demand

Chapter 5 stressed the disparities between value and price. While the concept of value is surely the more intriguing, the concept of price is the dominant theme in economic analysis. In reality, many questions of environmental economics center on items that have no specific market price. Sunsets, camping trips, and police protection provide comfortable examples. Yet, if the major problems associated with environmental quality are to be adequately understood, some input must come from economists.[1] Before their point of view can be appreciated, however, a few basic tools of economic analysis must be mastered. This chapter is intended to provide *enough* discussion of economic principles to carry the reader through the somewhat more complicated arguments of Chapters 8, 9, and 10. An interested reader can find expansion of these points in almost any current introductory economics text. The topics covered here include *only* the circularity of an economic system, the notion of marginality, demand, supply, and market equilibrium.

FEEDBACKS AND CLOSED SYSTEMS

Economics deals with mutually dependent support systems in which decision-making entities affect and are affected by each other. Producers require the existence of consumers and vice versa. It is helpful to think of this mutual relationship as a "feedback." Such a relationship is diagrammed in Figure 1. In the top panel, it is shown that a buyer has some influence on the seller's behavior, and the seller has some influence on the buyer's behavior. This is a closed feedback loop.

In real life not all loops are this tidy. A more complicated system might contain a buyer, a seller, and a third element, the government. This system is shown in the lower panel of Figure 1.

It is somewhat more complicated but does indicate that buyers affect sellers and the government. The government, through regulatory action, affects buyers and sellers, and the sellers, too, have an influence on the other two groups. An economy organized as in the lower panels of Figure 1 would be a complex one since a high degree of interdependence is present. All feedback loops are closed, however, indicating that when one group affects another, the latter group also has an influence on the former. An economy like that of the United States is a complex system requiring and thriving upon feedback and interdependence.

In a capitalistic system, two classes of decisions are always present and are always engaged in feedback exchanges. The family or household is one unit exerting its influence by making consumption decisions. The individual firm is the other unit; it makes production decisions. The two are related through straightforward feedback couplings.

MARGINALITY AND MARGINAL UTILITY

The workings of the feedback loops depend upon decisions made by each element in the system. Most individual decisions made in a market economy are termed *marginal decisions.* In economics, a marginal event is an added event. A sixth fishing pole added to a collection of five poles is a marginal pole. The decision to buy ten rather than nine loaves of bread makes the tenth loaf a marginal loaf. The last grade entered into a teacher's gradebook is a marginal grade. By definition, then, decisions are always made "at the margin." A marginal decision is made to buy the

A. A simple two-party exchange — each affects the other.

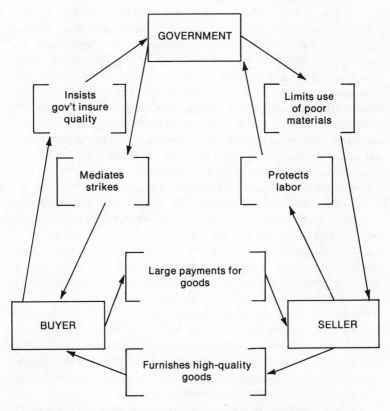

B. A three-party feedback system with each affecting the other two.

FIGURE 1: Two Examples of Feedback Mechanisms

first loaf of bread; a second marginal decision is required to obtain the second. This basic notion of marginality is the key to much of economic analysis. Decisions are made by producers and consumers on the basis of what contribution the marginal decision will make to the total satisfactions being enjoyed by the decision maker. In the case of a single consumer, the decision is always made with reference to the net added enjoyment that will be received by deciding to obtain an added unit of goods. The added enjoyment is termed the marginal utility of the added unit of goods.

If a single consumer purchases a single unit of a good, the marginal utility of that single unit is likely to be quite high. More and more units add successively less and less to the consumer's total utility or total satisfaction. The example shown in Figure 2 will help clarify this notion. When only a single unit of an item is consumed, the consumer receives eight units of satisfaction, or, the marginal utility of the first unit is eight. As the second, third, and fourth units are purchased and consumed, the marginal utilities drop from eight to seven (for the second unit), six (for the third unit), and five (for the fourth unit). In simple terms, as one acquires more and more green apples, the satisfaction derived from the additional apples diminishes. This basic principle regarding marginal utility can be used to explain why water, a necessity of life, is so inexpensive while diamonds, having no life-giving or life-sustaining qualities at all, are very expensive.

Since water is a necessity of life, the marginal utilities of the first units of water are so high as to make consumers willing to pay extremely high prices for them. But in most places water is plentiful; so much is available for consumption that its marginal utility is very low. Consumers are unwilling to pay more than a very low price. Diamonds present a different circumstance. They are expensive baubles and can be shown to be expensive using the same line of reason. Even though diamonds are not necessary to life, people want them and are willing to pay high prices for them because the marginal utilities of the first (and in this case only) units are very high. If more diamonds were made available, their ability to yield satisfaction would drop; the consumers' willingness to pay high prices would also diminish.

In the parlance of the economist, these two circumstances and the many possibilities between these extremes describe the law of *diminishing marginal utility*. The first unit of any product is very valuable because it has a high utility or a high capacity to satisfy desires and wants. As more units are obtained, the want-satisfying capacity, or utility, diminishes.

Another example may be found in the natural world. Many groups concerned with preserving wildlife have exerted considerable pressure on legislative bodies to protect endangered species of birds and animals such as the California condor. Detractors, attempting to counter the "endangered species" reasoning, have argued that anyone in favor of saving the condor from extinction must also be against killing any animal and must, therefore, be a vegetarian. This argument might sound plausible, but it is similar to the diamonds and water paradox. The California condor, because it is scarce and near extinction, has a very high marginal utility and should, therefore, be preserved. Domestic animals raised to provide food, hides, and fiber are, by comparison, very plentiful. Because the marginal utility of a condor (or brown pelican or peregrine falcon) is so high, one need only be an economist to want to preserve the species.

Figure 2 shows marginal utility and can be used to demonstrate the law of diminishing marginal utility regardless of the good in question. The graph can also be used to show the total

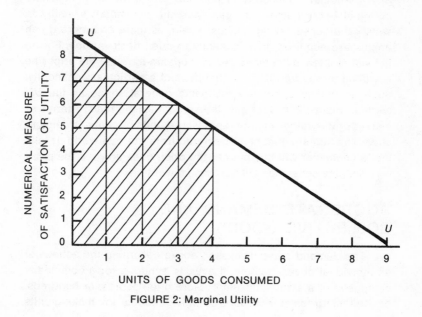

FIGURE 2: Marginal Utility

utility accumulating when additional units of a good are added to an existing collection. If the marginal utility of the first unit is eight and the marginal utility of the second unit is seven, the total utility derived from two units is fifteen (8 + 7 = 15). Similarly, the total utility of four units is 26 and is shown by the hatched area in Figure 2.

The law of diminishing marginal utility results in a curve represented by the heavy line, *UU,* in Figure 2. This curve shows that when only small quantities are available, they provide high utility to the users. When large quantities are consumed, only small amounts of utility attach to *added* units. This discussion is a useful introduction to the concept of demand.

DEMAND

Demand is a technical term in economics and has special meaning. A given household or person earns income by selling the factors of production it owns to producing firms or to another household.[2] The income earned is then used to purchase the necessities, desirables, and luxuries of life. These items are purchased because of their ability to satisfy the consumer. As more of a particular commodity is purchased, the law of diminishing marginal utility comes into play, and the commodity's ability to satisfy the consumer diminishes. Again, as more and more green apples are acquired, their capacity to quell desire drops. For an individual, then, a "demand curve" appears to relate the prices he is willing to pay for different quantities of a particular commodity. Such a demand curve has the same general shape as the *UU* curve in Figure 2, except now the vertical axis is read in terms of market prices rather than in the obscure units of utility. A demand curve has numerical content gathered from the real world. It tells that a consumer will buy more if the price goes down. If the price is raised, the consumer will purchase fewer units.

AGGREGATE DEMAND AND COLLECTIVE GOODS

The demand curve discussed above described the actions of an individual buyer. Seldom, if ever, is a market for a commodity composed of a single consumer. More often scores, or hundreds, or even hundreds of thousands of individual decision-making units are involved in the demand for a product. Hence, the simplistic version above, while describing individual behavior quite well, does not give an adequate picture of the whole market (all individuals interested in purchasing the same commodity). What is needed to explain a whole market is an aggregation, or summation of the behavior of all individuals participating in that market. Such a summation is carried out *horizontally*; that is, the price axis remains unchanged, but the quantity axis now refers not to what one person would buy at each price but to what all participants in a single market would buy. Again referring to Figure 2, the vertical axis now measures prices in dollars and since the graph deals with many buyers, the horizontal axis might be measured in thousands of units. At a market price of seven dollars per unit, 2000 units would be sold. At a lower price of five dollars per unit, 4000 units would be sold. The old curve depicting decreasing marginal utility is now called the market demand curve and provides an important building block in following paragraphs.

The horizontal summation process is shown in detail in Figure 3, where the separate demand curves for two individuals, *A* and *B,* are summed in the upper right panel. At a price of four dollars, each individual purchases four units. Hence, in the whole two-consumer market, a price of four dollars will yield sales of eight units. If the price is dropped to two dollars, each consumer will purchase six units and the whole market (again limited to two consumers) will purchase twelve units. The resulting aggregated demand schedule produces the traditional downward-sloping demand curve so prominent in the literature of economics. Note carefully that the demand curve used to describe the aggregate behavior of all consumers participating in a market for an ordinary commodity is derived through the horizontal summation of demand curves of individuals. In practice, it is usually not possible to construct a demand curve for a single individual, but many economists have isolated the demand curve for groups of people or for the entire economy. A caution is in order. A demand curve can be ascertained, but it is relevant only for a given income distribution, a given set of prices of other goods, given tastes, and given time periods. If any of these should change, the consumer's disposition toward the good in question will also change, and a whole new demand curve must be constructed.

The lower right panel of Figure 3 shows a *vertical* summation of demand curves. Vertical summation is used when the good being bought and sold has the special properties of a "collective good," which is a good that can be used simultaneously by more than a single consumer. Before vertical summation can be explained, a digression on property rights is needed to clarify the nature of a collective good.

A modern decentralized economy is based upon specialization and exchange. When exchange takes place, rights to the ownership and use of goods and services are transferred from one person (group) to another person (group). In the United States, a very large percentage of all material goods are privately owned, and their owners have exclusive rights to use them. This means that others can be prevented from enjoying the goods. When one person buys a lakeside cottage, all others can be excluded from

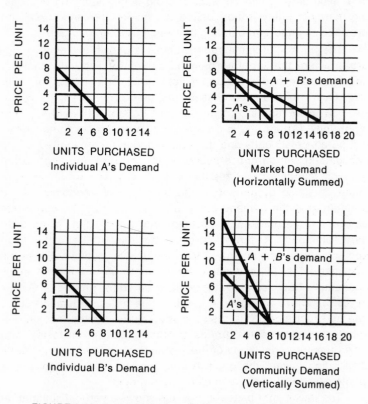

FIGURE 3: Vertical and Horizontal Summation of Demand Curves

the cottage. Or, when one person buys a loaf of bread, all others can be kept from eating the bread. Property rights are important in economics because some goods and services are produced so as to make exclusive rights virtually impossible to arrange. Radio signals are of this type. One person, two people, two hundred people, or two million people can simultaneously enjoy the same radio signals, rendering infeasible the establishment of a "market" where property rights to radio signals can be bought and sold. Exclusive rights to radio signals cannot be purchased. Goods displaying this characteristic are called collective goods. Much more will be said on this subject (see Chapter 9) since collective goods are often the key to issues in problems of environmental quality.

The vertical summation of individual demand curves (lower right panel of Figure 3) is needed to obtain a "market demand" or aggregate demand curve for collective goods. An example is provided by considering a public service such as police protection. A single individual may have a very ordinary looking demand curve for this service. He may be willing to pay a high price for the first units of protection but lower and lower prices for successive units. But police protection is not the kind of good for which exclusive rights are sold. Rather, it is provided on a collective basis. This makes sense since protection is provided for all citizens simultaneously. Because exclusive property rights for the services of policemen cannot be established, the usual horizontal summation of demand curves (which asks how many units the entire market will absorb at a given price) is no longer useful. The appropriate demand schedule is determined by asking what you and each of your neighbors would pay for the same unit of the good, then summing these amounts to find what any given unit of police protection is worth to the community as a whole.

Figure 3 is reformulated in the lower right panel to consider such collective goods as police protection. Each individual, A and B, still has his own demand curve for this service. As before, the demand curve for the individual relates to the price he is willing to pay for various quantities of the service. Since the service can be enjoyed simultaneously by A and B, the community demand curve adds the amounts that each would be willing to pay and a new vertically-summed demand curve appears. In the diagram, A and B would each be willing to pay four dollars for four units of police protection. The four units they purchase would be the same four units, of course, so the whole community (*A* and *B*) would be willing to pay eight dollars.

When dealing with ordinary goods, aggregate demand curves reveal how many units will be purchased at a given price. With this special class called collective goods, the curves show how much the community as a whole will pay to purchase a given quantity. In the former case, individual demand curves are summed horizontally; in the latter, they are summed vertically. This importance of this distinction will be apparent in Chapter 9.

A final modification must be added to the discussion of demand. In Figure 4 an aggregate demand curve for a particular commodity is shown as *DD*. The market price is three dollars per unit, so the public purchases slightly fewer than 5000 units. In the case of marginal utility, the area under the curve represents total utility. The area under a demand curve represents total willingness to pay. In Figure 4 the consumers of the (say) 4800 units would be willing to pay amounts equal to the entire hatched area in order to obtain the 4800 units. It would be impossible to ask each prospective buyer how much he is willing to pay, so the sellers charge everyone three dollars. Thus, revenues to the sellers (or payments made by buyers) are represented by the rectangular hatched area. A cross-hatched triangular area is left, representing the difference between what consumers would be willing to pay and what they actually pay. This triangular area is designated *consumers' surplus*.

COSTS OF PRODUCTION

Analogous to the law of diminishing marginal utility is the law of diminishing returns. While the law of diminishing marginal utility describes behavior in consumption, the law of *diminishing returns* describes a technical problem of production. It says that as the number of units of variable inputs (such as labor, fuel, or metal) added to a fixed input (land or a machine) increases, output will increase, but by successively smaller amounts. For example, an acre of wheat-producing land may yield 20 bushels of wheat when one man tills it. A second man, regardless of how hard he works, cannot eke a second 20 bushels from the land. His efforts bring total output up to only 35 bushels; he is increasing the yield by only fifteen bushels. Adding a third laborer may result in only ten additional bushels, bringing the total yield up to 45 bushels. The operation of the law of diminishing returns causes each worker to contribute less than previous workers to total product. Ultimately, this law prevents the world's food supply from being grown in a flower pot. It also gives rise to the calculation of a set of "cost curves" appropriate to an individual producing firm.

The costs of production in an individual firm are a complicated matter. They involve *fixed costs* that remain the same regardless of whether a single unit, 100 units, or one million units of output are produced. Important fixed costs are taxes, depreciation, and managers' salaries. A second class consists of *variable costs* which increase as output increases. Variable costs include labor, raw materials, and fuel; they are purchased only if the firm is in production. Finally, there is *marginal cost* made up of all *added* costs necessary to increase production by a single unit. These three concepts of cost — fixed costs, variable costs, and marginal cost — are diagrammed in Figure 5. Fixed costs and variable costs are shown as averages (per unit) and are summed (vertically) to give total cost per unit of output. Average fixed costs start very high, since at low levels of output the whole of the

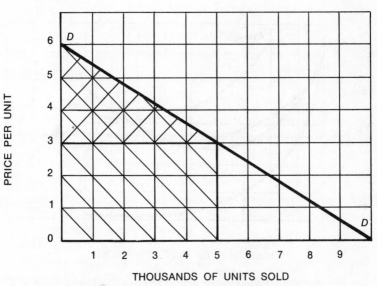

FIGURE 4: A Market Demand Curve with Modifications

fixed costs must be distributed among only a few units of output. Fixed costs per unit, however, drop rapidly as output expands. By 8000 units of production in this example, the fixed costs have been distributed among so many units of output as to be almost negligible. Further increases in output lead to further decreases in average fixed costs (but they never quite reach zero).

Variable costs are another matter. In order to produce the first unit of output, high variable costs must be incurred since many inputs such as labor and some kinds of materials must be purchased in large lots. But when output expands, the average variable costs decrease as the firm becomes more efficient. Eventually, the law of diminishing returns does its work, and average variable costs begin to go up. Average variable costs increase because, beyond some level of output, machines must be run too fast, managers lose control and become ineffective, and bottlenecks develop. The vertical sum of these two cost curves (average fixed and average variable) yields the average total cost curve, which also has a "U" shape.

Most confusing but also most important is the marginal cost curve. Generally, the marginal cost curve is very high when output is low but drops quite rapidly as output increases. Then, as with

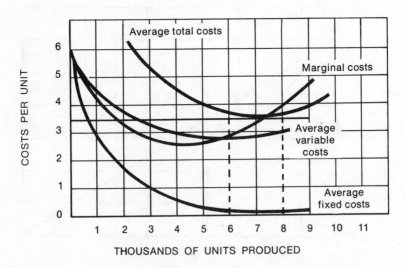

FIGURE 5: Relationships Among a Firm's Cost Curves Given a Specific Size of Fixed Plant

the average cost curve (and for the same reasons), the marginal cost curve begins to turn upward. The marginal cost curve intersects the average variable cost and average total cost curves at their lowest points. This *must* happen. If marginal cost were below average cost (whether average variable or average total), average cost would decline. If marginal cost were above average cost, average cost would increase.

The above discussion describes very briefly the relationships among cost curves. Included with the other costs of production is a set of costs called *normal profits*. Normal profit refers to the profit a businessman could earn if he were to invest his funds in some alternative venture. If a businessman is not earning normal profits in his present activity, some other field would be more profitable to him. Normal profits are included in the economist's cost curves because they represent the opportunity cost of investment — what the businessman could get if he invested in some other business.

A basic assumption in economics is that every decision-making unit strives to make marginal costs equal to marginal returns. Marginal return is the revenue associated with the sale of the last unit of output. For a firm so small as to have a negligible effect on the whole market (as with a single grower of potatoes or a shrimp fisherman), the price received for its product is its marginal revenue. The equation of marginal cost with marginal revenue makes good sense. Figure 6 shows that at a price (or marginal revenue) of two dollars, the producing firm should place about 5500 units of output on the market. If fewer units are offered for sale, the firm sacrifices profits since the returns on each of these units are less than the cost of producing them. If more than 5500 are sold, the firm loses money since the costs of producing the additional units are greater than the revenues they produce. The firm should offer more units only if the price goes up (Price II). This means, of course, that marginal revenue will have increased to the level of marginal cost.

This last feature bears closer examination. As product price changes, the most profitable or optimal output of the firm must change in order to keep marginal cost equal to marginal returns. The marginal cost curve denotes how many units of output the firm will place on the market at any given price. It thereby becomes the firm's *supply curve*. For a supply curve to accurately reflect behavior, all else must remain constant. Any alteration in time, place, or circumstances could cause the firm to alter its methods of production, and a new marginal cost and supply curve would emerge.

The supply curve derived from Figure 6 is for a single firm. Finding the supply curve for an industry is more difficult. An industry supply curve must consider that as output expands, many firms will begin to bid against each other for sometimes scarce factors of production. When this bidding takes place, the price of the factor changes and so, then, must the firm's marginal cost (supply) curve. A sophisticated industry supply curve would result only from a complicated technical study. But a first approximation may be attained from the horizontal summation of the supply curves of the individual firms. Figure 7 shows a simple supply curve for an industry.

MARKET PRICE AND EQUILIBRIUM

When demand and supply curves are used together, a market price emerges, and with it the notion of equilibrium. In Figure 8 the demand curve (the aggregate behavior of a group of consumers) and the supply curve (the aggregate behavior of a group of producers) intersect at point N. This point has as its coordinates a price of about five dollars and a quantity of about 7000 units. Given this relationship between price and quantity, all of the product placed on the market by producers is purchased by consumers. At a price above $5.00, say $5.90, suppliers want to supply a quantity (8800 units) in excess of that desired by consumers (4900 units). In such a situation, suppliers can bring their wishes into line with the wishes of consumers only by dropping the price.[3] At any price lower than $5.00, the quantity demanded will exceed the quantity supplied. Again, the resulting scarcity of the good is removed by consumers bidding against each other and raising the price back to the level N. This is the *equilibrium price.* Given these supply and demand curves, it is the only price at which there is no net force toward change. It is the only price simultaneously satisfying both buyers and sellers.

Understanding the relationship between the equilibrium price for an industry (as depicted in Figure 8) and the behavior of a

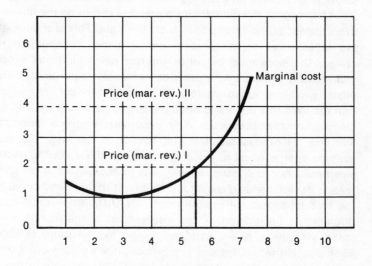

FIGURE 6: The Equation of Marginal Cost with Marginal Revenue

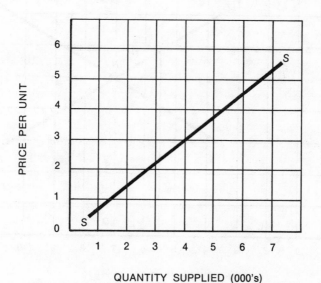

FIGURE 7: A Typical Industry Supply Curve

single firm requires an even more complicated diagram. In Figure 9 the behavior of the *industry* (or market) is shown on the left, and the behavior of the *firm* is shown on the right. This is a competitive industry, so no single firm can influence the price it receives. The price must be set by the free play of industry supply and industry demand. At a price of three dollars per unit, the industry produces approximately 3500 units of output. The single firm receives this price and attempts to equate this price with the marginal costs of production. This occurs at 35 units of output. At this level of output, the firm just covers all its costs. Because average cost is equal to price, the firm is neither making excessive profits nor is it losing money. It is in equilibrium. If for some reason demand should expand — that is, if the demand curve for the product should shift to the right — to D_1D_1, a new equilibrium price would be defined by the intersection of the new demand curve with the supply curve and a new market price of about $3.50 per unit would appear. The higher price would make it profitable for the firm to expand output until — as the rule insists — marginal cost equals marginal revenue. This equality occurs at

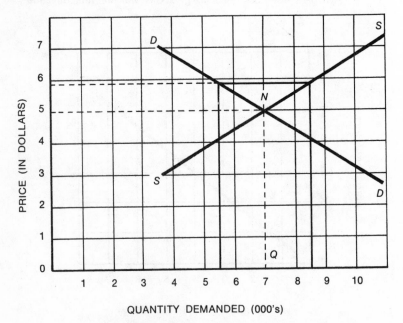

FIGURE 8: Supply, Demand, and Equilibrium

about 42 units of output and yields a situation in which price is above average cost. In other words, the firm is earning profits in excess of its normal profits. When this condition exists, the industry is not in equilibrium and, because of the profits being earned by the firms, output expands not only through firms already in the industry but through new firms entering. The new firms cause output to expand, shown by a shift in the industry supply curve to S_1S_1, restoring equilibrium at the old price and driving profits down. Equilibrium of the firm is restored when price is again equal to average cost.

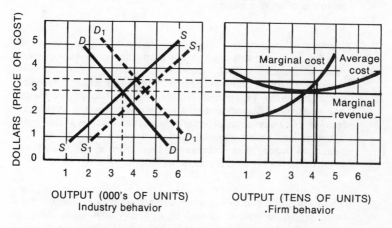

FIGURE 9: Relationship Between Firm and Industry

The brevity of this chapter has precluded a complete discussion of supply and demand. But a basic appreciation of these tools will help the reader to understand the more sophisticated issues raised in the following chapters.

FOOTNOTES

[1]Not only should a brief look at some economic principles help in understanding the problems, but it should also help in showing why traditional economic analysis cannot come to grips with some important qualitative problems.

[2]"Factors of production" are things (labor, machines, natural resources, etc.) that are used in the production process. For many years economists divided factors into three broad classes: land, labor, and capital. Some economists like to add a fourth: management. Others have regrouped factors into only two groups: human and nonhuman. Classifications are scarcely important. What is at issue is that households ultimately possess the factors (labor, management, and steam shovels), and earn incomes by selling their factors to firms. These incomes are then spent by the households to purchase the finished products of the firms.

[3]Some will argue that this is a naive treatment of the question since suppliers could expand demand by extensive advertising aimed at shifting the demand curve to the right. While true in the strictest sense, this is an unnecessary frill for present purposes. The point is that something must give if equilibrium is to be reestablished.

APPENDIX A: A Reformulation of the Diamond and Water Paradox

Early in Chapter 6 the question of the relative values of diamonds and water was discussed in terms of marginal utility. The same notion can be used to explain some of the frustrations being experienced by persons concerned with environmental quality. In the last two centuries, especially, the developed nations have produced an increasingly impressive array of goods and services for their constituents. These goods and services have accumulated to such an extent in upper middle class homes that the marginal utility of "things" has surely dropped. At the same time, fresh air, clean water, wilderness, and quiet (things often called "amenities") have one by one become more scarce, more dear. They now have rising marginal utilities. The curves in Figure A may help clarify a useful point. The curves show the decreasing marginal utility of goods over time and the increasing marginal utility of amenities over time. These changes in marginal utilities reflect not basic changes in values held by people, but changes in the availability of the two classes of objects. In the past, amenities — especially those associated with the natural world — were plentiful. The homesteader in Montana had all the clean air and quiet he could use. As a result, the marginal utility he received from amenities of this kind was quite low. At the same time, goods available to him were few in number — perhaps limited to cloth, tools, and a wagon. The relative scarcity of goods gave them high marginal utilities which in turn encouraged production of more goods and services. The low marginal utilities attaching to amenities allowed them to slip away almost unnoticed. As time passed, the relative marginal utilities changed, with the MU of goods steadily dropping while the MU of amenities steadily increased.

Marginal utilities are subjective and can never be given numerical value. But some people are now beginning to wonder if the decreasing marginal utilities of added goods and services are not roughly equal to the increasing marginal utilities of diminishing amenities. If so, the wisdom of producing more goods should be questioned since, by moving into the future, past time period P in the figure, the added utilities of goods and services would be lower than the marginal utilities of the amenities that had to be sacrificed in order to produce the goods. At point P', amenities with marginal values of OD have been sacrificed to obtain goods with marginal utilities of OA. The difference (BC) indicates again the loss attaching to the production of goods and services.

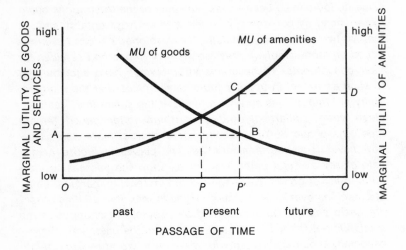

FIGURE A: Marginal Utilities of Goods and Amenities

APPENDIX B: Perfect Competition

In the body of Chapter 6, mention was made of the possibility of a firm so small as to have little or no effect on the price it received for its product. In a technical sense this is one of the requirements of firms in a perfectly competitive industry. A competitive industry is not the same to an economist as it is to the layman. The latter thinks competition among firms is demonstrated by aggressive salesmanship, advertising, and rivalry in business. Thus, auto dealers in a city are in competition, as are banks and super markets.

To the economist, on the other hand, a competitive industry is one characterized by many firms, no one of which can influence the price of the product they are all producing. Moreover, a competitive industry is open. Firms are free to enter or leave at any time; knowledge in the industry is available to all firms; and factors of production can move freely from one firm to another. Few industries are perfectly competitive. Most economists think some segments of agriculture come closest to meeting the requirements. Perhaps some segments of the commercial fishing industry do, also.

It is common for the economist's notion of perfect competition to be described as a norm or a goal for all society. Some argue that all production should be carried out under conditions of perfect competition. This is surely folly since, if all firms were to be forced to be so small as to have no influence on market price, they would not be able to take advantage of "economies of scale." Many goods would then have to be produced and sold at prices considerably above present levels. The perfectly competitive system remains firmly entrenched in economic textbooks as a pedagogical device showing the operation of a simple — not necessarily ideal — economic system. It will be mentioned occasionally later in this book.

chapter seven

A Nonmarket Tool–
Benefit/Cost Analysis

When goods and commodities flow through markets and have their prices determined by the free play of the forces of supply and demand, private decisions made by hundreds of individuals can adequately direct resources into production processes. Moreover, producers can make marginal adjustments in resource commitments in response to changes in factor or product prices. There is a large class of goods and services, however, for which no markets exist. Some other mechanism is used to decide if resources should be committed to the production of these goods. Other than those goods freely produced in nature, the most important goods and services that do not flow through market channels are those provided by government. Markets in the sense described in the previous chapter will not answer questions about how many resources should be devoted to schools, fire departments, highways, or flood control. Such questions must ultimately be answered through political rather than economic channels. Yet, politicians often turn to economists for advice and assistance. In order to properly advise political leaders, economists have had to develop procedures for evaluating nonmarket activities. One such procedure is benefit/cost analysis. Because of its prominence in natural resource studies, and because many attempts have been made to use it in other applications, this chapter will provide an introduction to this tool.[1] Economists and others seeking to study environmental problems should find it a very useful technique.

Because benefit/cost analysis is a combination of economic analysis and administrative decisions, it is quite easy to misunderstand some of its major features. The product of benefit/cost analysis is a ratio that can be used to determine if certain economic objectives will be met by a proposed program. The most common application of this technique has been in the water resource field where the benefit/cost ratio has been used to determine the appropriateness of certain (usually large-scale) investments. In these cases, a ratio of benefits to costs of less than 1:1 indicates that the benefits are less than the costs and that the proposed investment would lose money. On the other hand, a ratio in excess of 1:1 indicates the reverse, that benefits are greater than costs or that the proposed investment will return a "profit" (and contribute to increasing the well-being of the nation).

In symbolic form the benefit/cost ratio is generally expressed as $\frac{B}{K+O}$, where B equals present value of a stream of annual benefits,[2] K equals dollars of immediate investment, and O equals present value of all operation and maintenance costs that will be required through the investment's life. The terms require individual discussion.

The benefits (B) in the numerator of the fraction include the dollar value of net increases in production caused by a particular investment. For example, the benefits accruing to an irrigation project would be the sum of the values of all new agricultural crops whose production is attributed to the project. These items (such as apples, potatoes, sugar beets, and alfalfa) are likely to have market prices, so the benefits are merely the physical quantities of diverse goods multiplied by their respective prices. The products are then summed.

It is important to note that in this process market prices are used to build up to a tool that can be used in cases where no market value exists. For example, the market price of apples is used ultimately to obtain a "market price" for a public irrigation system. The problem is somewhat more complex in the case of flood control where there is no marketable output such as irrigated crops. In this instance, the numerator of the benefit/cost fraction is calculated with reference to the costs saved as a result of the availability of the protection. If spring flooding consistently does $500 damage to the front porch of a particular house, the provision of flood control would have as part of its benefits the $500 porch that now need not be replaced each year. Moreover, this $500 is counted as a benefit each year that the flood control activity is in operation. Each of the other major objectives of publicly-sponsored water investments (such as hydropower, navigation, municipal water, and recreation) has its own peculiarities of benefit measurements, but these two examples provide sufficient insight into the process by which a value is attached to the numerator in a benefit/cost analysis.[3]

One other peculiarity on the benefit side deserves mention. The benefits accumulated and placed in the numerator must represent changes that would not take place without the investment. A large water resource investment (or highway if you prefer) may hasten development of an area, but it is quite likely that some development would occur in the area even without the major public project. In Figure 1 the path *AB* designates no growth in dol-

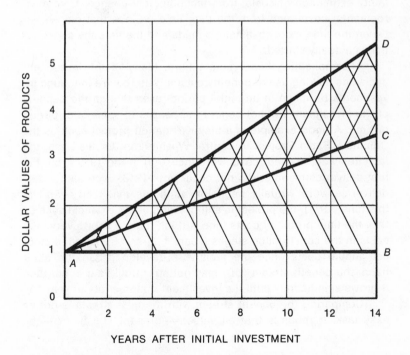

FIGURE 1: The Benefits Included in Benefit/Cost Analysis

lar output over a fourteen-year period. Growth path *AC* designates the growth that is expected if no investment in a public project is made and the growth path *AD* is the anticipated growth with the public project. It is proper to credit the project only with the benefits (those dollar values of goods) represented in the cross-hatched triangle *ADC*. These are the new benefits accruing because of the existence of the facility.

On the cost side must be placed the initial investment, interest on this investment, and the long stream of operation and maintenance costs anticipated over the life of the facility. If the investment is to add to the well-being of society, of course, benefits must be in excess of the sum of these costs. In most cases, the estimation of costs is not quite so hazardous as the estimation of benefits. Engineering data on dams and highways and parks are sufficiently refined so that decision makers need only present lists of the required amounts of land, concrete, steel, labor, and other necessary materials. Calculation of the cost side of the ratio is then virtually complete. Some difficulties always arise from faulty estimation of costs, from technological change, from environmental costs, and from the long time periods that elapse between the time a cost estimate is made and the time the construction is actually started.

A complicating factor enters here. Many of the investment decisions making use of benefit/cost analysis require that huge investments be made in an initial period, even though the benefits of the investment will not begin to appear until some later period of time. A good example is a major irrigation project such as the Columbia Basin Project in central Washington. In this case, the facilities were started (costs were incurred) in the 1930's, but the first deliveries of water did not occur until 1948. Even then, water deliveries were virtually negligible until the mid-1950's. And at this time (1971), the project is still growing. In an example such as this, the time profile of costs (line *AB*) in Figure 2 shows very high costs in the construction stages (years 1, 2, and 3) but quite modest annual costs once the facility is developed. On the other hand, the benefit stream (*CD*) may not start until year 3, but then it grows significantly until the investment matures (about year 15).

Comparing the benefit stream with the cost stream is not an easy task. It requires that all values — present and future — be put into a common frame of reference called "present value." The determination of present value is accomplished through use of the discount rate, an important tool to use in allowing for the passage of time.[4] A digression into the workings of the discount rate and its relationship to the passage of time seems warranted.

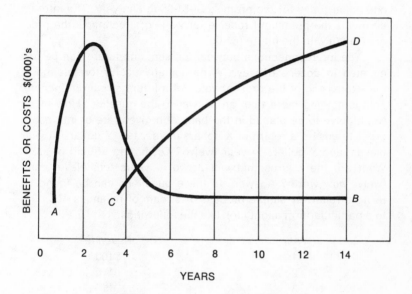

FIGURE 2: Time Paths of Benefits and Costs

Economic decisions would be quite simple if all were made at the same instant in time and if none had any lasting effects. However, many decisions by producers and consumers are made with long periods of time in mind. A consumer purchases a house with a 25-year mortgage. This causes effects extending over a long period of time. A large manufacturer decides to invest in a major plant at a new location. This, too, has effects that will be felt for many years. Similarly, the federal government may contemplate a dam on a major river. The dam may have an expected life of 50 years or even a century. Before making the huge outlays necessary for such a structure, the decision-making authorities and agencies must anticipate future events.

Economists have, since their earliest efforts two centuries ago, made connections between time periods through the use of the rate of interest. One formulation or explanation indicates that the rate of interest connects time periods by rewarding people for waiting. If the correct rate of interest is five percent, and I choose to put one dollar in the bank at this rate, I am giving up one dollar today for the promise of $1.05 in one year. The interest earned is the premium I receive for waiting; it connects the present ($1.00) with the future ($1.05).

The interconnections between present and future can be formulated in dozens of ways. A particularly useful formulation is the expression of the present value of a future stream of income. This is nothing more than an answer to the question: How much would have to be placed in the bank at a given rate of interest in order to ensure a return of X dollars in year ten, Y dollars in year eleven, and Z dollars in year twelve? Or, turned around, it asks: What are the returns that will arrive in years ten, eleven, and twelve *now* worth? Answers to these questions can be found by using a relatively simple process. A stream of revenues accruing to a particular firm might look like the following:

Year	Revenue
1	$100
2	110
3	115
4	125
5	150

The problem is finding the present value of $110 two years hence, $115 three years hence, and similarly for later years. A discounting formula can be used in which PV equals present value, R_t equals revenue accruing in time period t, and i equals the rate of discount.

$$PV = \Sigma \frac{R_t}{(1 + i)^t}$$

If the appropriate rate of discount is five percent per annum, the discounted stream of revenues will be as shown in the right-hand column of Table 1.[5] The sum of the annual present values is $514.72 — considerably less than the undiscounted sum of $600. The questions have been appropriately answered. It is known that if the rate of discount is five percent, $514.72 in the bank today would yield the above stream of income.

Not only benefits but also costs of long-term investments are discounted back to the present. This at first may seem strange, but a moment's reflection should clear the issue. Recall that in a large-scale, publicly-financed natural resource development a decision is made at one point in time. That decision commits funds to be used for special purposes for decades to come. Once the decision was made to develop Yellowstone National Park, the government was committed to pay not only the initial acquisition and development costs but also the operation, maintenance, and repair costs in the future. It is proper, then, to ask how much money would have to be invested at interest today in order to meet the annual costs as they arise. The answer is found by discounting the stream of costs in a manner similar to that mentioned above. The result is a benefit/cost ratio calculated with reference to the present values of the income stream and the cost stream.

Table 1
**Discounted Future Revenue
Assuming Five Percent Rate of Discount**

YEAR	REVENUE	$(1 + i)^t$	PRESENT VALUE
1	$100	1.0500	$ 95.25
2	110	1.1025	99.77
3	115	1.1576	99.34
4	125	1.2155	102.84
5	150	1.2763	117.53
Total	$600		$514.72

Severe and sometimes acrimonious controversies swirl about the determination of the "correct" rate of discount. There is no one "correct" rate, and rational men will continue to disagree for legitimate reasons. In determining which discount rate to use in a particular case, factors that must be considered include the market rate of interest, risk, price fluctuations, length of life of the investment, and source of funds. Such problems require more sophisticated analysis than is appropriate here.

However, selection of the rate of discount is a critical process and something must be said about it. Determining the values of benefits and costs accruing 25, 50, or 100 years into the future is a formidable task, but selecting the correct rate (or rates) to use in discounting is even more crucial to the adequate use of this decision-making criterion.

Figure 3 demonstrates the problem. Assume that the investment in question yields a constant benefit of $10,000 each year for 25 years. If benefits were not discounted to allow for the passage of time, total benefits in the numerator of the benefit/cost ratio would be 25 x $10,000 or $250,000, and could be honestly represented by the entire *ABCD* rectangle in Figure 3. If these same annual benefits of $10,000 were discounted at four percent, only the hatched area *ABED* would be counted. And if the discount rate were eight percent, only benefits equal to cross-hatched *ABFD* would enter the ratio. Similar results would be obtained on the cost side.[6]

A final theoretical problem must be discussed. Economic decisions, if they are to be optimal, must be based on costs and returns associated with the last unit of output produced or consumed. This is the "marginal" rule described in Chapter 6. If the marginal benefit of an additional unit of output (its contribution

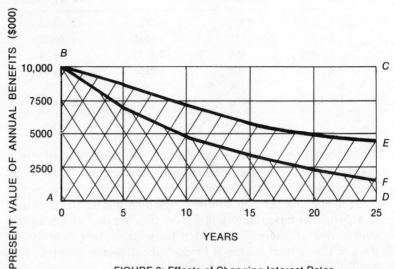

FIGURE 3: Effects of Changing Interest Rates

to total benefit) is greater than its marginal cost (its contribution to total cost), then output should be expanded since the benefit contributed by the expansion will exceed the cost of expansion. Production or consumption should be extended to the point at which marginal cost is just equal to marginal benefit. If it is not, some opportunities to increase profits or satisfactions will be lost. With respect to production, then, efficient decisions result when marginal cost is made equal to marginal benefits.

These same rules should also apply in government decision-making processes regarding natural resources and/or the environment. A major problem is the fact that the benefit/cost ratio is an *average* rather than a *marginal* concept. Figure 4 illustrates the problem that this can cause. Again, recall that a benefit/cost ratio of 1:1 or greater indicates an investment that adds to the welfare of the nation. Reading along the average benefit/cost ratio line in the figure, it can be seen that any investment between size *A* (about 1.6) and size *C* (about 5.3) meets this criterion. It is equally important to note that after the average curve reaches its highest level in point *B'*, the benefit/cost ratios associated with *marginal* increments in size drop quite rapidly. So rapidly in fact, that beyond point *B* (about size 4) each marginal unit of investment has a benefit/cost ratio below the 1:1 needed to ensure increases in welfare. Put in the language of the previous paragraphs this means that if a particular investment's contribution to national well-being were to be maximized, its size would be extended to *B*. However, given the fact that benefit/cost ratios are calculated as averages, a standard benefit/cost calculation would appear to show that any investment level between *A* and *C* is acceptable.

Despite serious problems such as those mentioned above, benefit/cost analysis in one form or another has been used for decades (or perhaps centuries) by governments in making decisions regarding the allocation of resources. Among its many practical advantages is its simplicity. Most anyone can visualize what is taking place when simple comparisons are made between benefits and costs. Most anyone can easily understand that a ratio greater than unity means a positive return and a ratio less than unity yields losses to the system. This is one of the major reasons why benefit/cost analysis was chosen as the official criterion to be used in many public decision-making arenas. In the United States, the choice was formally made in 1936 when Congress stipulated in the Flood Control Act that, if a federal water development project is to be authorized and built, benefits "to whomsoever they accrue" must be in excess of estimated costs. A large number of rules and practices have emerged from this general concept.

No pretensions need be harbored about the relationship between economics and politics. Most decisions in the area of natural resource development have to be public decisions. Such problems as flood control and irrigation are most often too large to solve using private sources of funds and private decision-making processes. Public decisions, simply because they are public, involve politicians and are sometimes suspect on these grounds. Many have argued — some persuasively — that decisions in the natural resources field have been more political than economic. This view need not be contested nor need it be dwelled upon since, as in most instances of this type, evidence can be gathered to support a variety of conclusions. The point is that political decision makers at the federal level have consistently attempted to incorporate some economics into their information-gathering scheme while retaining a good deal of noneconomic

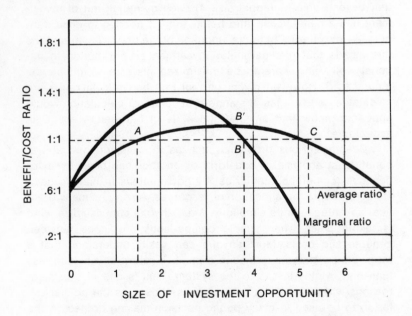

FIGURE 4: Marginal and Average Benefit/Cost Ratios

power over the decision itself. Thus, a host of expedients have emerged. The first of these expedients was the development of a four-stage decision-making process which includes the following:

1. Initiation: A local area brings to the attention of its congressional delegation a problem worthy of consideration. If a problem does in fact exist, a project is proposed to solve it.

2. Justification (determination of feasibility): An intensive engineering and economic study is ordered to determine if the proposed project is physically and economically sound. This is the step during which formal benefit/cost analyses are conducted.

3. Authorization: Congress places its stamp of approval on the project, effectively telling the planning group that, if they can find the money, they may proceed.

4. Appropriation: Funds are voted to develop the project. The appropriation is usually part of a larger bill, but occasionally a project is of sufficient importance to be handled separately.[7]

Only the second stage has significant economic content, and the economics involved in that step is greatly attenuated. The test of feasibility asks only one economic question: Will the project return more than it cost? All projects receiving an affirmative answer to this question are placed on equal footing in the political processes that follow, namely the authorization and appropriation steps. A project with a benefit/cost ratio of 1.01:1 is as legally acceptable as one with a ratio of 9.2:1. This legal peculiarity can effectively be described as an expedient — it includes the use of some economics but permits considerable political manipulation in deciding which natural resource endeavors to fund. No attempt is required beyond the separation of the good projects from the bad projects. No priorities are established among the good projects based on economic principles.

The above discussion of benefit/cost analysis is only a brief introduction to the complex world of nonmarket decision-making devices. It seems to underscore the fact that when the market does not work, elaborate devices must be used if economic decisions are to be made. Benefit/cost applications provide an opportunity to see what has historically been done to make economic decisions in the absence of market data. Smog, water pollution, noise, and vanishing amenities are problems that cannot be analyzed within the ordinary workings of the market. If these problems are to have solutions, devices similar to benefit/cost analysis will surely have to be used.

FOOTNOTES

[1]Extensive literature on many facets of the benefit/cost issue can be found in: Otto Eckstein, *Water Resource Development* (Cambridge: Harvard University Press, 1958); Roland McKean, *Efficiency in Government Through Systems Analysis* (New York: John Wiley & Sons, 1958); Robert Haveman, *Water Resource Investment and the Public Interest* (Nashville: Vanderbilt University Press, 1966); and S. C. Smith and E. N. Castle, eds., *Economics and Public Policy in Water Development* (Ames: Iowa State University Press, 1964).

[2]The concept of present value is discussed later in this chapter.

[3]Not all the goods and services of publicly sponsored natural resource projects can have monetary values placed on them. Boating on the lake behind a flood control dam, for example, is very hard to "price." Some of the difficulties of nonpecuniary values are discussed in Chapter 12.

[4]"Discount rate" is a technical term referring to the process by which future incomes or sums of money are converted (discounted) into present value. In a strict sense, the rate of interest and the rate of discount need not be the same. An intriguing exposition of the problems which surround determination of the "correct" rate of discount can be found in W. J. Baumol, "On the Social Rate of Discount," *American Economic Review*, vol. 58 (September, 1968), pp. 788 - 802.

[5]Most often, it is not necessary to go through all the mechanics of discounting. The process is used so frequently that bankers, loan agencies, and others who deal with money and time use prepared tables to find the present value of dollars in future time periods. One set of these tables can be found in E. B. Cox, *Basic Tables in Business and Economics*, (New York: The McGraw-Hill Book Company, 1967).

[6]As noted before, the "correct" rate of discount is an unsettled issue. Some economists argue that high rates should be used in discounting benefits accruing to natural resource projects. Others argue that very low rates should be used. This issue need not become a complicating factor, but it must be mentioned that the use

of low rates means benefits accruing in the distant future have a higher present value than if high rates were used. Low rates of discount are often thought of as being more compatible with conservation since they put high premiums on things that will be available in the future. To point out the irony associated with the interest rate controversy, it can also be said that high rates of discount are compatible with conservation since they prevent development of many kinds of resources!

[7]There are literally scores of steps which must be taken within each of these four stages. A rigorous review procedure plus much political maneuvering through the elaborate congressional committee system is part of the process. A brief view of the process can be found in R. H. Haveman, *Water Resources and the Public Interest* (Nashville: Vanderbilt University Press, 1965), pp. 13 - 21.

chapter eight

Market Failure-Externalities

INTRODUCTION

The market system can be regarded as a vast and complex social machine that organizes the economic activities of individuals and firms. Like all machines, the market system is highly specialized. Some functions it performs well, some not so well, and some not at all. Unfortunately, matters of environmental quality fall mainly into the latter two categories. This chapter will explore one of the more important reasons for the failure of the system to adequately deal with environmental problems. In the tradition of economic theory and literature, this reason falls under the heading "market failure."[1]

EXTERNAL EFFECTS

One particularly upsetting failure of a market-oriented economic system is commonly known as an external effect or externality. There are several other names for these effects. They can as well be called uncompensated effects, third-party effects, or simply non-market effects. External effects and externalities will be used here.

In every economic transaction, one party incurs costs in order to receive benefits. The other party receives payments and gives up goods or services. This demonstrates interdependence and closes the feedback loop. In an ordinary transaction, the purchaser is expected to pay the full cost of the item, and he expects to get full and sole claim to its use. For example, a homemaker who buys a loaf of bread expects to pay the full cost and to get the full enjoyment from consuming it. Unfortunately, the feedback loops in economics do not always close this completely or this surely. Sometimes those who pay the costs do not receive all the benefits; and sometimes the payments made for an item do not cover all the costs of producing it. An example of the first type is the man who sprays his swampy backyard so mosquitos will no longer breed there. His neighbors receive some benefits even though they have not contributed any payments. An example of the latter condition is the factory that dumps raw industrial waste into a nearby stream. It is not paying all the costs of its doing business.

In either of these instances, the system may be described as being inefficient. In the first case (the payer not receiving all the benefits), it is likely that not enough of the product is being produced. If each of the people benefiting from the mosquito spraying program had been assessed for his share of the benefits, more funds would have been available to spray other yards. When such assessments are not or cannot be made, the system does not produce enough of the product or service (mosquito abatement). In the second case, the system may be producing too much of the product. If the factory had to pay all the damages created through its polluting activities, its cost of production would rise, and its products would have to sell at higher prices.

Many of the most common problems of environmental quality can be traced to some variation of the problem of external effects. Air and water pollution stem from the fact that automobile owners, industrialists, and municipalities can avoid many costs associated with waste disposal by simply dumping their refuse into the atmosphere or into waterways. Lumbering companies can cut trees without having to pay those who prefer to leave the trees standing and who, therefore, suffer a loss when forests are cut. On the other hand, no single individual wishes to pay for clean air because he would have to clean up all the air for everyone. Everyone will share in the benefits of his purchase, but few will be induced to help him pay for it.

All external effects have two properties: interdependency — one person's behavior creates a cost or benefit to other persons; and lack of compensation — the one who creates cost is not made to pay for it, nor is the one who creates the benefit completely rewarded for it.

With this in mind, the desired relations of exchange and the unintended relations of externalities can be shown. A producer and a consumer, regulate each other through the supply and demand mechanisms of the market, as shown by the arrows of influence in Figure 1. But the producer creates other effects, smoke and noise, affecting the welfare of a group of citizens who do not consume the product of the firm. There is no corresponding arrow of influence through which these citizens can regulate the producer's effects. Therefore, the producer and consumer will reach an equilibrium position different from that which would be optimal were all affected citizens' interests taken into account.

A realistic and more equitable solution to this problem could be achieved through a government agency receiving complaints from the affected citizens and transforming these inputs into regulatory intervention (taxes, subsidies, laws, or regulations) into the

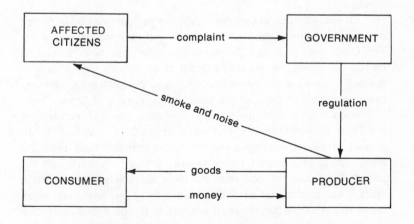

FIGURE 1: Exchange, Externalities, and Closing Feedback Loops

producer's affairs. The feedback loop would then be closed through the path affected citizens ⟶ government ⟶ producer and a more nearly optimal state would be attained. The producer is now forced to include the interests of those who suffer from smoke and noise in his decision-making process.

The above is much too abstract. It was earlier noted that externalities take two forms — external costs and external benefits. Each is sufficiently important to the question of environmental quality and control to warrant separate and more specific treatment.

EXTERNAL COSTS

External costs can be illustrated through use of an exaggerated but somewhat familiar situation. For this purpose, International Noxious, a fictitious but large industrial conglomerate, will be studied. International Noxious (IN) has located one of its factories on the banks of the river Clear, a large stream flowing through the metropolitan area of Bliss. A coalition of the local Chamber of Commerce, the realtors' board, and other civic boosters was able to lure IN into the area by offering the firm a package deal including property tax exemptions, tax-free municipal bond issues to finance the factory, and, not least, the propinquity of the Clear which provided a natural and obvious drain for IN sewage.

Previous to this economic coup, the residents of Bliss fished and swam in the waters of the Clear and sported on its banks. The Clear also provided the municipal water supply of Bliss and its few industries, not least important of which was the nationally-famous brewery established (according to the brewery's own testaments) solely to capture the unique advantages of Clear water.

Soon after the IN plant began operation, the fish catch from the Clear dropped sharply. The water of the Clear was further observed to cause rash and nausea upon contact. Now only the most daring and expert water skiers ventured on its broad expanse. Publicans noticed that the local beer acquired a certain "bite," not unlike that of overly ripe kidney pie and that, when taken in sufficient quantities, it created euphoria closely paralleling that of uremic poisoning. (The beer is, however, much in demand for this property, and the market is growing.)

The citizens of Bliss appear to have adapted to the altered condition of the Clear. A recent study has shown that fully 90 percent of those who previously fished the Clear now spend an average of $20 per month journeying 150 miles to fish at Reject — the area that lost in the struggle for the IN installation. These expenditures have significantly increased regional (and national) income.

While the residents of Bliss have recently voted down a two million dollar bond issue to provide parks and swimming facilities for the residents, sharp observers predict that only modest opposition will develop for the $250 million issue to purchase water from Reject and transport it across the Rising Gorge mountains to meet Bliss's expanding industrial and municipal needs for pure, unpolluted water. The Country Club has also announced plans to add nine more holes to its 18-hole golf course to accommodate the increase in membership drawn from IN executives and various local realtors. These plans, however, are contingent upon obtaining water rights from the new diversion project.

Given this general background on the situation in Bliss, the economic implications of external effects may be shown more exactly. The hypothetical cost and demand curves for pollution treatment by IN are shown in Figure 2.

Figure 2 shows two extreme positions on the horizontal axis. In Situation I, with only one unit of effort being used to purify the river, all effluent is being dumped into the Clear at nearly zero cost to IN itself. In Situation II, IN has been forced to treat its effluent to such an extent (nine units of effort) that the Clear is nearly pure. Neither situation is likely to be optimal for either IN or Bliss. As the degree of sewage treatment increases from Situation I and, consequently, the amount of sewage entering the river decreases, the marginal costs of sewage treatment rise. At the same time, the benefits of removing the last few pollutants may be small and becoming smaller. That is, the marginal benefits decrease as the pollution abatement procedures increase in intensity. There must be an optimal amount of sewage treatment lying somewhere between "nothing" and "all." There is an optimal amount of pollution just as there is an optimal amount of anything else. This optimum would rarely, if ever, be at the "no pollution" level.

The curve *MB* shows a conceptual measure of the marginal benefit to Bliss of increasing pollution treatment. When effluent discharge is very high, the marginal benefit of reducing discharge by some small amount may be quite high as shown on the left-hand side of the figure. As the discharge drops, the stream becomes increasingly pure and the added benefits accruing from further attempts to reduce effluent may be quite low. Put another way, the benefits attaching to efforts to clean up an already clean stream will certainly be very low, while even modest efforts to clean up a badly polluted stream may yield very high benefits.

The curve *MC* measures the marginal cost to IN of sewage treatment. This marginal cost curve shows that, when large amount of pollutants are being introduced into a stream, the cost of eliminating a few of these is likely quite low; but as the firm is required to reach increasingly severe standards of purity, marginal costs rise. The cost of getting rid of the last few pollutants will be quite high. The result is a marginal cost curve that rises rapidly as purity (Situation II) is approached.

It is important to note that, in this simplified case, costs are being borne by IN, while benefits are being received by all the

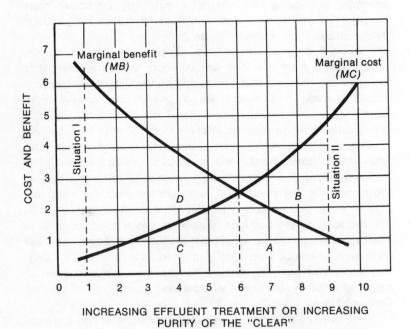

INCREASING EFFLUENT TREATMENT OR INCREASING
PURITY OF THE "CLEAR"

FIGURE 2: Benefits and Costs of Purifying the "Clear"

residents of Bliss. When six (6.0) units of effort are expended to treat effluent, the marginal benefits of sewage treatment to the people of Bliss equal the marginal costs of sewage treatment to IN. At all points to the left of six, benefits to Bliss are more than the cost to IN. At all points to the right of six units of effort, sewage treatment costs IN more than it benefits Bliss. The optimum amount of pollution control is, therefore, at six units rather than at full abatement (Situation II).

Pollution control programs can be instituted and carried out to the optimal point, with benefits to Bliss exceeding costs to IN. Total benefits equal to areas *C* + *D* are realized, while costs of only the magnitude of *C* are incurred. As control is extended beyond six units, additional benefits equal to area *A* can eventually be realized, but these are insufficient to pay the total costs beyond the optimal level of area *A* + *B*. It is inefficient from an economic point of view to extend control beyond 6.0. The question remains: Should benefits designated by area *A* be forgotten?

There is no easy answer. On the one hand, if Bliss as a community were asked to pay the costs of treatment beyond six units of effort (area *B* plus area *A*), it would likely refuse since benefits are much smaller than costs. On the other hand, the community did enjoy the benefits represented by area *A* before the IN installation. So it is unreasonable for Bliss to assume the costs of treatment beyond six units, and it is unreasonable as well as inefficient to require IN to treat sewage beyond this point.

In the end, the question of what to do about *A* reduces to a decision regarding what is fair and reasonable. Two solutions are available: (1) IN can be forced to reduce effluent to the six-unit level and then pay the amount represented by area *A* to the residents of Bliss in cash compensation for the remaining effluent damage; or (2) IN can be permitted to retain the amount *A* as a kind of reward for reducing pollution to the optimum level. Either solution results in the economic optimum 6.0 units of abatement. The question of how to handle area *A* is ultimately a value judgment.

The preceding discussion shows that there is an optimal amount of pollution control just as there is an optimal amount of other goods and services. The problem now is how to induce IN to adapt its operations to reach that optimum.[2] The solution requires a larger framework of analysis; it involves what is fair and reasonable, what is politically feasible, and what the information and control costs of alternative techniques are. It is one of the very severe and very broad problems of environmental economics. Each technique of controlling external costs must be evaluated in this larger framework. Two very important techniques of control are (1) tax-subsidy programs and (2) legal restraints.

The technique most generally favored by economists is to force a polluter, such as IN, to "internalize" the external costs associated with its waste products. Internalizing a cost is nothing more than making it a part of the firm's continuing cost structure, like labor and fuel. It is as if a tax had been imposed on the discharge of untreated wastes. If the tax were high enough, IN would be forced to reduce its wastes (possibly be relocating) rather than pay the tax; if the tax were low, IN might choose to pay the tax and continue polluting. Because of this latter possibility, taxing schemes are sometimes derisively described as providing "licenses to pollute." The same basic effect could be achieved by having the community of Bliss pay IN a subsidy to be used for cleaning up the factory's wastes.[3] Thus, Bliss could pay IN an amount equal to the height of the marginal benefit curve (*MB*) for successive amounts of effluent treatment.

One can legitimately quarrel about these alternative means even though the same outcome is attained. To the executives of IN, a tax appears to be a penalty imposed on them after they have already established their factory in Bliss. The company was probably not told that such a tax would ever be part of its manufacturing costs. A subsidy, on the other hand, smacks of a protection racket whereby the residents of Bliss are forced to pay to protect themselves from the threat of being polluted by a plant that many of the residents did not think would produce pollutants. On these ethical problems, reasonable men can disagree. There is no general rule to decide which is correct — the tax or the subsidy. But two suggestions can be made.

First, the *ex post facto* problem of the tax can be alleviated (or certainly mollified) by giving the industry sufficient time to adapt to new circumstances. The capital investment made under the old circumstances will depreciate over a certain length of time, and the new capital (equipment) that replaces it can be designed to reduce pollution costs. The best way to ensure that industry will in fact design these pollution-free features into their plants is to confront them with the prospect of a *future* pollution tax. This is a way of saying to an industry that they will not be penalized for past actions, but present actions will not be tolerated in the future.

The second suggestion is based on the premise that, to a point, the social gain from pollution control (area *D* of Figure 2) is quite large while the costs are modest. It is not necessary that either the residents of Bliss should receive all the gains, as in the case of a tax, or that IN should receive them all, as in the case of a subsidy. The gain can be shared.

A tax could be imposed on IN to force it to treat its effluent to the point that the marginal cost (with the tax) is equal to marginal benefit. Any added treatment deemed necessary by the residents of Bliss could then be paid for through a subsidy. The tax-plus-subsidy program is a means of cost-sharing between IN and Bliss. Another method currently employed is to give companies tax advantages for investing in pollution treatment equipment. These tax advantages lower the costs of pollution-free production and are, in effect, cost-sharing programs.

The most common type of pollution control program, however, is not a tax program, a subsidy program or a combination of the two. It is legal restriction. Under threat of heavy penalties (fines, the closing down of polluting firms, or even jail sentences), it is simply made illegal to pollute over a certain limit. This is an effective way to get the job done, but it is often not the most economical.

The optimal place to put the legal restraint is defined by the intersection of the marginal benefit curve and the marginal cost curve (six units of effort in Figure 2). If only one firm is producing all the pollutants (IN polluting Bliss), there is little difference between the legal restraint and a tax-subsidy alternative. When many different firms and many different municipalities are involved in polluting an area, certain advantages of the tax-subsidy technique become apparent. Different firms are likely to have different cost curves for effluent treatment. That is, the marginal cost curves for some firms will be higher than they will be for others. Each firm will, therefore, have its own optimal amount of treatment. Confronted with a tax schedule, each company will find its own optimum position. Those with high costs of treatment will treat less and pay more tax; those with lower costs will treat more and pay less tax. Thus, the job of cleaning up the river will be allocated to those who can do it most efficiently. If a uniform legal restraint is imposed on all firms, this kind of adjustment cannot be made. All will have to meet the same standard, so some will treat too little of their effluent and some too much.

Of course, if there is a sufficiently small number of polluters, it may be possible to assign each its own legal restraint in terms of its own optimum. Then the above objection can be overcome. The choice between tax-subsidy programs and legal restraints ultimately reduces to a question of the information and control costs of each and cannot be settled on an *a priori* basis.

There are certain inherent advantages to a legal restraint. In terms of public pressure and legislative understanding, it may be easier to obtain legal restraints because these approaches avoid complex public revenue questions. Legal restraints can also be made to provide a basis for private suits against violators and may thereby become somewhat self-policing. In sum, legal restraints may not be as good as a perfect tax-subsidy program, but they are often the only alternative to nothing at all. Nor does the choice necessarily have to be between these two techniques. Indeed, the best way of regulating pollution may be by a combination of both — a legal restraint prohibiting pollution over a certain amount and a marginal tax on any pollution less than that amount. In the last analysis, legal restraints are actually extremely high taxes.

To return to Bliss and its problems, the regulation of IN effluent, whether by tax-subsidy programs or by legal restraints, provides the essential feedback mechanism to close the loops connecting the citizens of Bliss with the activities of International Noxious. Figure 3 is an adaptation of Figure 1 showing this specific result. The damaged third parties, who are suffering from the effects of IN effluent, now pressure governmental agencies to do something about the damage. These agencies, in turn, relay back to IN measures designed to correct the ill. The loop is closed.

But a significant problem remains. With any control short of complete subsidy, IN will have higher costs of production after regulation than before it. The installation of elaborate antipollution devices will cause profits to fall, output to drop, or both. How do these events affect Bliss and society as a whole?

As shown in the appendix to this chapter, the answer depends upon the degree to which the regulation of IN conforms to the regulation of industry as a whole. If the citizens of Bliss force IN to incur the costs of sewage treatment, but all of IN's competitors producing in other locations are permitted to continue polluting streams without similar constraints, then IN will be forced into a position of earning less than normal returns on its investment and will eventually be forced out of business. The costs of production are now higher for IN than for its competitors. Local regulation could cause substantial losses of employment and income in the community of Bliss. Indeed, in monetary terms these losses could far outweigh any gains of sewage abatement for the community as a whole.

If, on the other hand, the regulation affects all producers in the same fashion, the result is quite different. The supply curves for all producers shift to the left after the fashion described in Chapter 6. Prices of consumer goods rise and a new competitive equilibrium is attained, with all firms again receiving a normal return on their investment. Indeed, under these circumstances IN may support its own regulation out of a sense of "social responsibility" or for purposes of community relations. But it could not unilaterally clean up its own sewage without suffering a competitive disadvantage.

It must be recalled that in a competitive world consumers will have to pay higher prices for commodities when regulation of sewage disposal activities is present. This is a price they would be more than happy to pay as long as regulation brings with it net social benefits. Thus, the real income of society as a whole would increase. There may be, however, a quarrel over the dis-

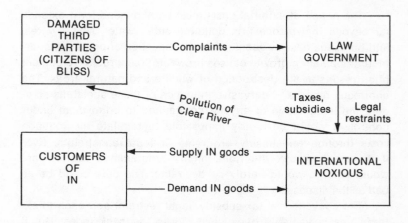

FIGURE 3: Feedback and Externality Again

tribution of this increased income. The residents of Bliss would be the main beneficiaries, whereas the costs would be borne by all consumers of IN products, regardless of where they live. But, again, if all industries were regulated in this way, the distribution of the costs and benefits would tend to balance out.

The essential principle is this: Regulation of polluting industries must be universal. It must be applied equally to all firms in a system. Piecemeal regulation of one firm at a time can result in losses that far outweigh the benefits, not only to the producing firms but also to local areas and eventually to society as a whole. This problem is common in economics; it will be encountered frequently in the pages to come. Some processes are simply not additive. In complex systems the whole is not merely the sum of its parts. Indeed, systems have a kaleidoscopic property which means that small alternations among some parts can create wholly different patterns in the system. As has been seen in this elementary example, regulation of IN's industrial waste problem can be either good or bad for society as a whole. It depends upon how it is handled.

EXTERNAL COSTS AND SEPARATE FACILITIES

The notion of external costs must be expanded for it extends far beyond that of one firm polluting air or water. Noise, wires, and buildings destroy scenic vistas; congestion cripples transportation; population growth causes crowding; resource development often results in the destruction of wildlife and natural areas. The enormous range and pervasive influence of these social ills staggers the imagination of anyone determined to bring them under control. It would be nearly impossible to regulate all pollution costs through tax-subsidy programs or legal constraints. Even if it were possible, the rigid, highly programmed society that would emerge would hardly be desirable. The cure could be as bad as the disease.

The conventional tax-subsidy, legal restraint approach to externalities is valuable because it closes feedback loops. But it really does not get at the heart of the problem. It does not reveal why these loops apparently tend to open and become increasingly costly as society develops. Externalities of the kind faced by modern societies are, in many respects, modern problems. Primitive societies were able to avoid or control them. Why this is so deserves some brief comment.

One reason is that primitive societies lived in a style described by Kenneth Boulding as a "cowboy economy." Waste products in these societies could be thrown over the fence. Even when the accumulation of waste rose to uncomfortable levels, the tribe could simply move to a new, virgin habitat (allowing the old one to recover). Now, however, people live on the other side of that fence, and waste thrown there lands in their laps. The accumulation of social and physical impediments in modern life makes it impossible to move, and the supply of new places to move to is rapidly disappearing. Human beings and human activities have effectively filled the world habitat; there is "no exit."

In more formal terms, pollution control in modern economies is so difficult because the marginal cost of pollution tends to rise much faster than the actual physical rate of pollution. Pollution costs are functions of frequency and density. Modern pollutants accumulate over time. Unlike the common cold, they do not simply go away. Social and economic forces bring people into ever closer interactions with each other, and the environment becomes progressively dense, whether in terms of particles or people. The frequency with which any given individual is affected greatly increases, and costs rise accordingly. But, while people are being brought into increasing interdependence, the degree and scope of face-to-face contact is rapidly declining. In more closely knit societies, everyone knows who made all the racket last night. In modern society this is not known. Consequently, automatic mechanisms of social pressure and personal responsibility within a small community are now disappearing and are being replaced with formal controls that are inadequate to the assigned task.

In sum, as the environment is progressively filled with human activities, the marginal costs of externalities rise dramatically. At the same time, past social mechanisms of control are disappearing. It is a bleak picture. And since all these effects are intimately related to the explosive forces of population and economic growth, the prospects for the future are even worse.

The root cause is interdependence in social and economic activities combined with isolation in social responsibility. At least a partial solution to the problem may lie in reversing these trends. Ezra J. Mishan has suggested an interesting technique. He proposes to reduce the degree of interdependency by creating "separate facilities."[4]

According to Mishan, special towns should be built where no automobile transportation is permitted and where no aircraft are permitted to fly overhead. People who value peace and quiet would have the option of choosing that way of life in a facility separate from the ordinary city. He also thinks that beaches and other recreational areas should be segregated to meet diverse tastes. Thus, part of the beach should be given over to the beer and transistor radio crowd, part to the nature lovers, and part to families. Such an arrangement could result in a much more efficient allocation of resources than shared facilities.

This line of thought has some precedent. The remote areas of the United States officially designated "wilderness areas" are a form of separate facilities made available through public action. Communities such as Pebble Beach, California, where the wealthy have gathered to enjoy separate facilities are clear examples in the private domain. The "commune" can be the poor man's equivalent of Pebble Beach. In a sense, separate facilities are a partial reversion to Boulding's cowboy economy; they are a way of keeping disagreeable people and things "over the fence" and are one of the most promising means of controlling rapidly spiraling external costs.

EXTERNAL BENEFITS

External benefits, like external costs, are unintended results of economic choices. The analysis of external benefits is similar to that of external costs. A producer or consumer creates a benefit for a third party. The creator of the external benefit bears the full cost of his activity but does not receive full compensation. Hence, from society's point of view, he tends to produce too little of the activity in question.

A familiar example of an external benefit occurs in hydroelectric power generation. In order to drive electric turbines on a year-round basis, it is necessary to construct storage reservoirs that hold heavy spring run-offs for release during the remainder of the year. This alteration in stream flow creates external effects on persons residing both downstream and upstream. Many of the external effects are negative or harmful, but substantial positive effects, too, are created. Floods are mitigated; maintenance of a stable stream flow enhances fishing and navigation activities; and the temporal reallocation of water from wet spring months to dry summer months may create substantial irrigation benefits to farmers. If the power producing company is not rewarded for these external benefits, it maximizes profits only with reference to power generation objectives. The external benefits will, therefore, be shortchanged, and many potentials for increasing social benefits may be overlooked.

Important examples of external effects are also found in consumption. A familiar instance is in education. Without a continual input of educated people into the social system, scientific advance would soon cease, industry would stagnate, and social and political institutions would likely decay. Yet, only a portion of these benefits (scientific and cultural advances) accrue to the individual student. If he were forced to bear the full costs of his education, he would purchase only the amount that would equate his personal marginal costs to his personal marginal benefits. This could be an insufficient amount of education from the point of view of the whole society.

In Figure 4, D_I is a hypothetical demand curve for education of an individual student (and/or his family). The marginal cost of each year of education is shown by the marginal cost curve. It is seen that, if left to his own devices, the student would be willing to consume only about eleven years of education. At this point he would pay an amount (say $3800) for the last year of schooling. But assuming that there are many social benefits stemming from this student's education, the social demand curve D_S must be added *vertically* to the individual's demand curve D_I. The vertical distance between D_I and $D_I + D_S$ reflects the added benefits that accrue to society as a result of having an educated populace. The optimum consumption of education for the individual and society as a whole is fourteen years. The student, however, is willing to pay only $3100 for the last year of education. Society, if it is to receive the full benefit from education, must pick up the tab for the difference between the $3100 paid by the student and the $4100 marginal cost of the fourteenth year of schooling. This dif-

113

ference is the value of the external benefit. It is economically efficient to subsidize education by the amount of the external benefit. Failure to do so would result in an insufficient consumption of education from a social point of view.

Occasionally, external benefits and external costs are coupled in such a way that the imposition of the external costs destroys some external benefit, forcing society to bear double costs. For example, if a child has been using a nearby cattle watering pond for fishing, wading, and other recreational purposes, that child is receiving a benefit external to some neighbor's production of cattle. If the pond is destroyed as a side effect (external cost) of urban sprawl, the child is deprived of a recreational opportunity. The child suffers an external cost of sprawl, but the story does not end there. Society may also have an interest in the child and the child's use of the pond. Society receives some external benefits from the child's recreational activity. When the pond is sacrificed for urban development, there is a double loss — the loss borne by the child and the loss borne by society. Yet, in the typical instance, neither the direct benefits to the child nor the external benefits to society from that child's play are considered in the decision to "sprawl" or not.

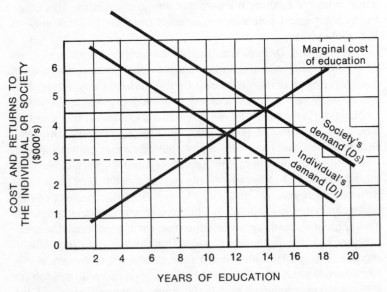

FIGURE 4: Individual and Social Demand for Education

EXTERNAL BENEFITS
OF THE ENVIRONMENT

*B > C of providing
it. Private people
won't
in many
cases.*

The example of a child's play in a pond introduces one of the most important yet least understood aspects of environmental economics. Does society receive external benefits from the use of the environment for aesthetic and recreational purposes? If it does, then certain policies such as public support and provision of outdoor recreational resources must follow. If it does not, outdoor recreation may be "a good just like any other good," best given over to the mercies of the market place. There are three basic arguments in support of the contention that outdoor recreation does create external benefits: (1) it serves a basic *need* of man; (2) it provides an *outlet* for antisocial impulses; and (3) it builds *character*. Each will be examined in turn.

1. *Need.* Thorstein Veblen asserted that one of the few universal properties of behavior is "the propensity for play among men and the higher animals." Veblen, indeed, reduced human progress to the working of this instinct for play. It stimulates the "idle curiosity" of man which, he believed, is responsible for the truly inventive developments in human history.

If it is true that man has a basic need for play, then it can be argued that one of the duties of a responsible society is to provide opportunities for this need to be satisfied. In this way public support of outdoor recreation can be defended on the same basis as public support of food, medicine, housing, and other essentials. There are, of course, other forms of play than outdoor recreation — spectator sports, casual reading, and courtship are but a few. But one important feature of play is that it must have a great deal of variety — many games must be available. Outdoor recreation provides a greater variety of activities than perhaps any other single kind of recreation.

Of course, the benefits of the natural environment are not limited to such active sports as fishing, hunting, and hiking. There is something to "communion with nature." The peace and tranquility, the stimulus to reflective thought provided by nature, are very real to anyone who has experienced them. A few days with nature provides something more than relaxation — it is like spiritual catharsis from which one emerges in some way reinvigorated. Why this is, or indeed what it is, is largely unknown. But it is a real phenomenon that may be of critical importance to man.

115

2. *Outlet.* The officials of Imperial Rome — some of the most capable public administrators in history — based their administration on two principles: "Bread and circuses." One of the current justifications for public provision of recreation in slum areas is that it provides residents with something to do. Certainly, public parks, swimming facilities, and other forms of outdoor recreation are considered an important weapon in the struggle over the urban crisis.

The basic argument of "outlet" can be (and has been) expanded to defend the more violent forms of recreation such as boxing, football, and hunting. Given a man-as-beast philosophy, or even a violent-man-because-of-modern-frustrations philosophy, these sports can be defended in terms like those used in defending the original Roman circuses. It is better for a few animals to die or for a few professional mercenaries to be maimed occasionally than to have these aggressive impulses turned loose on society in general. No one who has watched the audience at one of these modern circuses can reject this argument out of hand.

3. *Character.* Experiences during preadolescence and adolescence largely determine what kind of adult will emerge. Society has an interest in the kinds of citizens it will make for itself. The intimate connection between adolescence and play provides, therefore, a natural subject for concern.

Experiments in adolescent psychology have shown quite conclusively that the environment in which a child is raised substantially affects his ability to learn. Children raised in drab, unchanging surroundings tend to be dull. Those raised in a changing, variegated environment are more alert and curious. Curiosity and learning ability are acquired attributes.

In this connection, the phenomena of nature may be very important. Not only does nature provide a great variety of things to see, but these things are very peculiar. They stimulate curiosity because they are so hard to understand. Human activities are easy to understand at this level because the learner is himself human. Humans do things "because they want to." But nature provides its multiform variants for no clear reason and by mysterious means. It provides a strong stimulus to curiosity. The test of this hypothesis is easy: Take a small boy on a camping trip.

It is also argued that certain forms of recreation have a salutary effect on the developing child in a dimension other than intelligence. Certain forms of recreation, and particularly certain forms of outdoor recreation, are essentially family activities. Indeed, they are perhaps the last, vestigial remains of activities that can be so described.

A cause and consequence of modern urban civilization is the specialization of functions. This is a source of great efficiency; it is also a powerful impetus to isolation, alienation, and frustration. Specialization of functions has extended to the modern family. The father has his job to do, the mother hers, and the children go off to school. In the evening the parents watch their television programs, and the children watch theirs. (In poorer families they squabble over programs.) On weekends the father golfs, the mother does whatever mothers do, and the children, hopefully, play with other children.

On occasion, however, the family may take an outing together. When this occurs, the object is almost invariably some form of outdoor recreation. Thus, they may go camping or fishing, to the zoo, or to the museum. On these occasions all members of the family have a common interest. They are able to communicate intelligently, act together, and share their experiences. If society is interested in the maintenance of families, the "socialization" of children, and amiable relations between parents, it may derive very real external benefits from outdoor recreation.

A few other observations may be made in this connection. First, as a test, *one* parent may isolate himself, together with the children, in an environment other than the home. The isolated parent will soon find there is nothing to do except engage in some form of (outdoor) recreation. This experiment is amply confirmed by those who have had the normal family structure disrupted by divorce. When the children visit the outcast parent, there is little, other than outdoor recreation, they can do together. The experiment only brings into perspective what is generally the case in the normal family. But in the family, the lack of something to do together is hidden under the camouflage of institutional normalcy. It is no more than what is expected.

Second, the controversy over public provision of free recreational facilities centers mainly on family-oriented recreational activities. Thus, small game hunting, camping, stream fishing, zoos, museums, and parks are the objects of controversy. Few advocate free public provisions of skiing, big game hunting, deep sea fishing, or Himalayan expeditions. The evoked image of the family at play gives this issue its intense emotional coloring.

117

Finally, it may be noted that football, hunting, mountain climbing, and other competitive and dangerous sports are supported on grounds that they help develop that aggressive individualism and blind respect for the rules without which Western civilization would presumably perish. "The Battle of Waterloo was won on the playing fields of Eton" was a popular slogan. An irreverent wag replied, "Yes, and the Empire was lost in its classrooms." Whatever the desirability of this kind of character building, few who have gone through the monstrous phase of "growing up" can fail to appreciate the intimate relation between how one has played and what one has become.

In sum, it appears that the available evidence, though pitifully small, leads to the conclusion that, for a variety of reasons, society derives external benefits from various forms of outdoor recreation. This is particularly true of the recreation of children. Occasionally the hypothesis has been put to the test.

In San Francisco, recently, the electorate of that most urbane city was asked to approve a bond issue to provide recreational facilities in the Hunter's Point neighborhood, one of the city's poorest ghettos. The issue was approved by a majority, but it failed, however, to achieve the necessary two-thirds vote (it created an outcry about "one man, one vote"). The point is that the affluent majority were willing to tax themselves for the benefit of people they would never meet (nor, perhaps, care to meet) to provide recreational facilities they themselves would never use.

SUMMARY

In this chapter the problem of externalities has been discussed in some detail. The specific instances of externalities are numerous and the challenges they pose to a basically individualistic, free-market economy are profound. But externalities are not problems of capitalism alone. They afflict all societies and systems, whether they be socialist or tribal, urban or agrarian. Thus, the civilizations of the Mesopotamian Valley finally succumbed from within due to salinity in the soil. This condition was one of the external costs of irrigation. The Soviet Union today is afflicted with pollution problems of such magnitude that they, like us, are bordering on a national crisis. And, to add further irony to the problem, the Soviet Communists, together with the archetypal capitalists of Japan, are mainly responsible for the rapid extinction of the blue whale. There are no simple "revolutionary" answers to the problem of externalities. It affects all societies whatever their ideological persuasion.

A good deal more must be said about externalities. Several of the points raised in this chapter will have to be developed in more detail. But before going further, another type of market failure, "collective goods," will be introduced in the next chapter.

FOOTNOTES

[1]See Francis M. Bator, "The Anatomy of Market Failure," *Quarterly Journal of Economics,* vol. LXXII (1958) and William J. Baumol, *Welfare Economics and the Theory of the State,* rev. ed. (Cambridge: Harvard University Press, 1965).

[2]For a more rigorous and comprehensive treatment of this complex subject, see Allen V. Kneese and Blair T. Bower, *Managing Water Quality: Economics, Technology, Institutions,* published for Resources for the Future, Inc. (Baltimore: Johns Hopkins Press, 1968). See especially Chapters 5, 6, and 7.

[3]*Ibid.,* pp. 101 - 15, and the references cited therein. This is a controversial point in economic literature.

[4]Ezra J. Mishan, *The Costs of Economic Growth, op. cit.,* Chapter 8.

APPENDIX: Regulation of Pollution Through Taxation

This appendix follows the basic analysis of the perfectly competitive firm outlined in Appendix B, Chapter 6. It shows the probable effects of a pollution tax on IN and on the competitive industry of which IN is a member.

Figure B shows the hypothetical cost and price relationships of IN before and after a pollution tax is imposed (Figure A will be ignored for the moment).

Prior to the tax, all IN effluent is discharged into the Clear at nearly zero cost to the firm itself. The company is in equilibrium (producing at a point which equates marginal cost and marginal returns) and receiving a normal return on its investment. The relevant marginal cost curve is MC_1, and the average cost curve is AC_1, yielding an equilibrium output of OQ_1.

If the citizens of Bliss become outraged and force IN either to cease discharging sewage into the Clear or to pay a high tax on effluents, IN will respond in the manner shown in the preceding Chapter. It will treat effluents to the point where MC equals MR and pay taxes on additional polluting activities. The effect of this new requirement is to raise both the marginal and average costs of production. The IN plant now faces the curves AC_2 and MC_2. The rules do not change. IN still operates in a fashion that equates marginal cost with marginal revenue. Marginal costs are now higher, but, since IN has no influence on the market price, marginal revenue is still at level P_1 and output is reduced to quantity OQ_{II}.

At this output, the production of each unit of output brings with it a loss designated by the vertical difference between average cost and average revenue. This difference multiplied by the number of units produced yields the total loss sustained by the firm. The loss is shown by the hatched area, L.

The difference in these two situations is not that there is an economic loss in the second situation and no loss in the first; rather, the difference is one of who pays the loss and how the loss is felt by the community. In the second situation, IN is experiencing all the costs of regulating effluent discharge through hard cash outlays. In the first situation, the community in general bears these costs as they are diffused through the population in the more elusive form of lost opportunities and degradation of

Very good

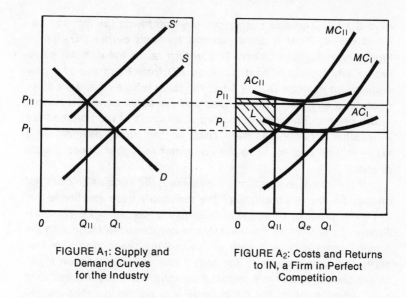

FIGURE A₁: Supply and
Demand Curves
for the Industry

FIGURE A₂: Costs and Returns
to IN, a Firm in Perfect
Competition

environmental quality. *Environmental losses borne by a community
are converted into monetary terms only when something has to be
done about them.*

A significant problem of IN's losses remains. As explained in
Chapter 6, IN will continue to sustain these losses as long as it
remains in business. But this situation, obviously, cannot last forever. A loss position is ultimately a disequilibrium position, and
forces will be set in motion to find a new equilibrium position.

In order to understand how the new equilibrium will arise, it
is helpful to make four simplifying assumptions which will subse-
quently be relaxed. These assumptions are: (1) that every firm in
the industry producing IN-type products is identical in terms of
internal cost and effluent discharge; (2) that the same marginal
tax is imposed on every firm in the industry; (3) that the tax is
not so severe as to force firms to shut down altogether; and (4)
that all effects occur in the "short run" — a period of time too
short to allow for the size or number of plants in the industry to
be expanded or contracted.

Under these conditions, IN and every other firm in the industry will reduce production from OQ_I to OQ_{II} of Figure B. This will
cause the industry supply curve of Figure A to shift to the left

121

from S to S'. The new supply curve S' intersects the industry demand curve, D, at a point yielding reduced output OQ_{II} but a higher price P_{II} than before. This in turn raises the demand curve facing any individual firm (such as IN) from P_I to P_{II}. The new equilibrium position Q_e will be at the point where IN and all other firms have eliminated their losses and are again earning a normal return on their investment. In a word, all costs of the pollution tax — whether effluent treatment costs or the direct payment of a tax — will be passed on to the consumer in higher prices for the products.

If the above assumptions are relaxed, the conclusion does not change (with the exception of the monopoly case mentioned below); the process of adjustment simply becomes more complex. Briefly, if the firms in the industry are not identical, those with the highest costs may be forced to discontinue production, this also shifts the supply curve to the left. The same may happen if different tax schedules are imposed on different firms. In the "long run," investment in the IN industry will be discouraged by the loss position, so no new plants will be built nor will present plants be expanded. Old plants will be allowed to depreciate away. These various effects will combine to produce the same result: the cost of pollution control will be passed on to the consumer, and a reduced but more nearly (socially) optimal amount of the product will be consumed.

Two points need emphasis: (1) If just one firm in the industry is taxed, this industry equilibrium will not occur, and IN will be forced to sustain losses if it remains in production at all. It is likely that ultimately it will be forced to cease production. (2) The analysis has assumed a perfectly competitive world. In situations of "imperfect competition" — in which IN can take steps to control the price received for the product — not all of the increased tax need be passed on to the consumer. A monopoly would absorb some of these taxes in the form of reductions in the excess profits they receive.[1]

FOOTNOTE

[1]Paul A. Samuelson, *Op. cit.*, Chapter 26.

chapter nine

Market Failure–
Collective Goods

Chapter 8 explained how the unintended consequences of economic activities can lead to overproduction of some things and underproduction of others. These consequences are called externalities, and they exist because the market fails to take them into account. This chapter deals with another market failure, *collective goods*. Most consumer goods and services in any economy are, or eventually become, private goods. Such private goods may be used by only one person at a time. Their use is competitive — either you use them *or* I use them. Some goods, however, are not competitive in use. Two or more people can simultaneously use them. Indeed, some of these goods may be used simultaneously by a large number of people without diminishing the supply of the good. They are called collective goods.[1]

A common example of a collective good is a lighthouse. Once it is operating, a great many ships can use the service of the lighthouse without in any way affecting the quantity or quality of warning available to the others. Rare species of plants and animals are also collective goods. The same number of whooping cranes can simultaneously satisfy the demands of large numbers of people who want the cranes preserved. Reception of radio and TV signals is a collective good since any number of persons can tune in without forcing anyone else to tune out. A scenic view can be shared by a great many people without diminishing the supply to any one individual. Within limits, the protection provided by national defense, police, and fire stations are collective goods, as are great works of art, the stock of knowledge in a society, and the national parks.

There are also collective "bads." A noxious odor can simultaneously nauseate any number of people in its vicinity. Similarly, any number can become ill because of poisons and diseases borne by the environment. An ugly building can disgust innumerable people. When a species becomes extinct, thousands mourn. Economic analysis can shed considerable light on items — good and bad — having these peculiar collective properties.

ECONOMICS OF COLLECTIVE GOODS

There are two basic kinds of collective goods: (1) goods such as lighthouses and national defense that are produced through technical processes and (2) goods that are part of the natural world — air, animal species, and forests. Regardless of the category into which a particular good falls, if it is scarce it must be evaluated and allocated in economic terms. While fish populations naturally reproduce themselves without direct cash outlays by man, economic decisions must still be made regarding whether to fish more now and have less later, or vice versa. The benefits of more now (or later) must be weighed against the opportunity costs of less later (or now). Whether created by man or nature, collective goods will be subject to economic decisions if they are *scarce* relative to the demand for them.

The economic theories and policies applicable to collective goods differ somewhat from those used to analyze private goods because the latter are competitive in use, while the former are shared. The remainder of this section will examine the economic implications of *sharing* — in the evaluation, allocation, and financing of collective goods.

The evaluation of a lighthouse will be used as an initial example. It is assumed for purposes of simplification that the lighthouse is a *pure* collective good. This is not strictly accurate because a point could eventually be reached at which the ocean could hold no more ships in the vicinity of the lighthouse, thus preventing unlimited sharing of the service. (In any case, the assumption is later relaxed, and congestion effects among users of collective goods are discussed.)

Total benefits attributable to the lighthouse are determined by the sum of the individual demand curves for every user. Demand curves for collective goods must be added vertically rather than horizontally to obtain total (aggregate) demand (following the reasoning developed in Chapter 6). Shipowner A can receive signals from the lighthouse without interfering with shipowner B; they simultaneously receive the same service. A third shipowner, C, or a fourth, D, can also simultaneously use the lighthouse. To obtain the total value of the services these ships are simultaneously receiving, the values placed on the signals by each separate user must be summed. To accomplish this, the demand curves are vertically summed. By contrast, the shipowners' demand curves for such private goods as fuel must be added horizontally to obtain total demand.

Total demand for the lighthouse (vertical summation) is shown in Figure 1. The horizontal axis shows the number of uses (warnings) received by any *single* shipowner. The vertical axis shows what the individual shipowner would be willing to pay for varying quantities of the service.[2] The vertical sum of the three demand curves (D_A, D_B, and D_C) is shown as the total demand curve. As explained in Chapter 6, the value of total benefits provided by the lighthouse is given by the area under the total demand curve, *TD*.

The total cost for operating the lighthouse is constant. It costs a given amount to build the facilities and another amount to man it. If the light is operating, total costs are the same regardless of whether one, ten, or 1000 warnings are received each hour.

If the area under the total demand curve *TD* (representing the total value of lighthouse services), exceeds the magnitude of the total costs, then total benefits are greater than total costs. The lighthouse has passed a major economic test. Whether or not it should be built depends upon other considerations, such as other uses for the "lighthouse money." But if the ratio of benefits to costs is greater for the lighthouse than for those alternative uses, the light should be built (other factors being equal).

A second question arises after the lighthouse has been built and is in operation: How should the services of the lighthouse be allocated among its users? The answer may be found in the constant nature of operating costs. As noted, costs do not vary with the amount of service provided. This demonstrates the dominant characteristic of collective goods: Many people can share them at no additional cost. This is the same as saying the marginal cost of a pure collective good is zero. Therefore, following the logic of marginal cost pricing of Chapter 6, the optimum price for a pure collective good is zero: Let everyone have all they want for free!

The reason for this is apparent. If a price of two dollars were charged, shipowner A would stop using the service, and shipowners B and C would reduce their use considerably. Total use would drop from nine units to under seven units. This would reduce the total benefits shipowners (and society) receive from the lighthouse by the area under *TD* between seven units and nine units of output. Yet total costs are the same at seven as at nine units. Therefore, society would experience a net loss in benefits by reducing use to any level below the maximum.

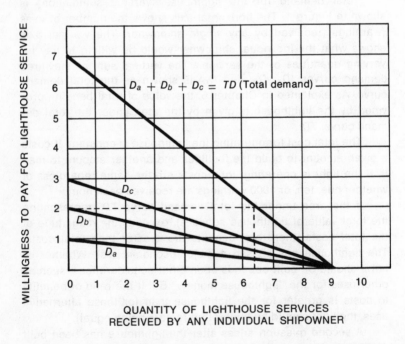

FIGURE 1: Individual and Total Demand for Lighthouse Services

If the optimal price charged for the service is zero, and if it costs money to provide the collective good produced by the lighthouse, problems of financing are bound to arise. Whoever establishes the light and charges the optimal (zero) price will have no revenue with which to pay the costs. The benefit/cost evaluation showed that the lighthouse would provide benefits in excess of costs and should, therefore, be constructed. That decision is independent of how the installation is to be financed. But once the decision to provide the lighthouse is made, the question of how to pay for it cannot be ignored.

Generally speaking, the best way to finance pure collective goods is through taxing each user. When the individual users can be identified, as in the case of a lighthouse, there should be a "lighthouse tax" imposed on that segment of the population. When, as in the case of clean air, the benefits are dispersed over the whole population, a general tax is warranted and should be levied. The economic advantages of a tax in financing pure collective goods are that (1) all beneficiaries must pay and (2) payment does not affect the amount used. If each shipowner has to pay $100 per year whether he uses the lighthouse or not, he would obviously use it to the full extent.

In sum, a pure collective good is *evaluated* through benefit/cost analysis using vertically-summed demand curves. It should be *allocated* at zero price so that all users will use it to the fullest extent possible. And it should be *financed* by taxes, either on individual users or on everyone who benefits, directly or indirectly, from the good. This completes the discussion of the efficiency aspects of collective goods, but it is not yet completely clear *why* the market mechanism is likely to fail in providing an optimal quantity of these goods. This question is the subject of the next section.

COLLECTIVE GOODS AND THE FREE RIDER

In order to explain the failure of the market to properly allocate resources to the production of collective goods, the lighthouse example may be reconsidered. Assume that individual shipowners A and B decide to form a consortium among all shipowners to build the lighthouse through private means. Two facts are apparent: (1) as the number of users increases, the net benefits of the lighthouse increase; and (2) as the number of shipowners contributing to the expense of the lighthouse increases, the average cost to everyone decreases. Both facts stem from the fixed-cost nature of the lighthouse. Therefore, it will be in the interest of everyone to have as many shipowners as possible join the consortium.

But will all shipowners join? It is very unlikely. Whether a shipowner joins or not, he will still be able to enjoy the benefits of the lighthouse if others build it. There is no way, short of using torpedo boats, for the members of the consortium (who have incurred the expense) to prevent non-members from enjoying the benefits of their efforts. Non-members can thereby obtain a "free ride" on the backs of members. Because of this potential opportunity for a free ride, it is unlikely that the lighthouse would be built, even though it would create substantial net benefits to society as a whole.

Each individual shipowner will be tempted not to join the lighthouse-building consortium hoping that "others" will build it. Each will want to get all the benefits from the lighthouse without having to pay any of the costs. Since everyone is playing the free-rider game, no one will actually build the lighthouse. And ships will continue to crash against the shoals.

The lighthouse has been used as an example because it represents a kind of collective good *least* subject to free-rider problems. The private interests involved (shipowners) are few in number, and the benefits and costs are obvious to all. Yet, even under these elementary conditions, the free-rider game can lead individuals to behave in an irrational manner — that is, against their own interests. The next example shows how this irrationality can work itself out in more complicated instances to constitute one of the great tragedies of our time.

MIXED GOODS: THE BLUE WHALE

Most goods are thought of as having one and only one use. A steak is for eating, an automobile is for transportation. This is rarely the case. Both steaks and automobiles are also very good for impressing one's neighbors. Some goods, unfortunately, are private in one use and collective in another. These *mixed goods* are often the focus of "conservationist" controversies.

An example is the sorry history of the blue whale. This magnificent beast is on the verge of extinction. He is disappearing because of his qualities as a private, commercial good. If the blue whale should happen to be saved, contrary to all expectation, it will be because of his collective good qualities.

As a private good, the whale can be captured, killed, and made to provide large quantities of valuable oil and meat. On the other hand, countless millions of people derive pleasure from simply knowing that the whale exists. The thought of these massive yet graceful creatures swimming through the deep creates a nearly mystical sensation. Without the whale, *Moby Dick* would never have been written. Collective good benefits flow from the sheer existence of such beasts, and the world would be very much the worse without them. This mixed (private-collective) property of whales is shown in Figure 2.

The worldwide demand for the existence of the whale as a collective good is given by the vertically-summed demand curve D_C. This curve cannot be derived through the use of objectively obtainable data so it must be estimated on the basis of a more

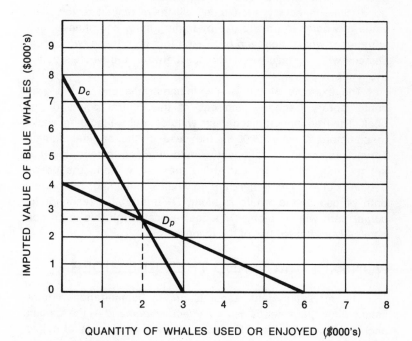

FIGURE 2: Private and Collective Demand for Blue Whales

subjective measure of willingness to pay. The market demand for the whale as a private good is considerably easier to determine and is shown by the horizontally summed demand curve D_P. Given the way Figure 2 is drawn, 2000 whales becomes an important number. If the whale population should increase beyond this level, the value of the added whales for collective existence purposes is much lower than their value for private, commercial uses (D_P is above D_C). If, however, the number of whales drops below 2000, their collective value greatly exceeds the private value (D_C is above D_P). The crucial 2000-whale-level is determined by the relative positions and relative slopes of the two demand curves. The private demand curve has only a modest slope since there are numerous substitutes for the meat and oil products of the whale as a private good. The steep collective (existence) demand curve results from the fact that there are no substitutes for the whale.

As time goes on, more books are written about whales, and more television time is devoted to them. More people are exposed to whales and learn to appreciate them as a collective good. Simultaneously, technology will have tended to lower the value of whales in private use by finding more and cheaper substitutes for whale meat and oil. Taken together, these two circumstances will increase the difference between the whale's collective and private value.

The example of the whale demonstrates one of the great contemporary tragedies and one of the central points of this book. The free-market mechanism will not cease allocating whales to commercial uses (2000 in the case of the original demand curves) as economic rationality would suggest. Instead, the market mechanism will tend to grind inexorably along the private demand curve D_P until there are no whales remaining. The *optimal* path of use of the whale is along D_P until D_C is reached; the *actual* path will be along D_P alone, even though this is an *economically inefficient* use of the resource.

NONEXCLUSION AND THE FREE RIDER

The market system responds not to humanitarian requirements or welfare needs but to *effective demand* — the amount people are willing and able to pay. Indeed, in the case of collective goods, the whole market mechanism works against a rational compromise that would be stated in terms of increased welfare.

The existence of various "Save the . . ." organizations, like the one endorsed by Joan Baez to save the blue whale, is testimony to this failure in the market.

Such organizations and campaigns encounter almost insuperable problems in trying to get their demand for a collective good translated into *effective* demand. There is virtually no way to organize consumers of collective goods so that they can directly bribe a whaler not to kill a blue whale. Even if the organization had the money to accomplish the bribe, the feedback loop between the fisherman and the organization would not close automatically or easily. These organizations must work through indirect channels. They must work for laws to preserve the whale or attempt to discourage the consumption of whale oil and whale meat. These are weak stratagems for closing the loop and saving the whale.

The problem is fundamental. Each individual who might like to preserve the whale faces strong inducements *not* to transform this interest into effective demand. While his demand for collective goods such as the whale may be very large, the effective demand created by his investment of time and money will be pitifully small.

There is an important reason for this aside from moral turpitude. The structure of an individualistic society provides strong incentives for rational economic man to conceal his demands for collective goods in the hope that he can satisfy them without having to make sacrifices. He may be able to get a free ride. If a person is asked to contribute time and money to an organization interested in preserving the blue whale, he may think the following: "The $50 that the whale is worth to me may be crucial. But if I give it, there is no assurance that the whale will be saved; if I withhold it, the contribution of Joan Baez and others like her may save it anyway. Therefore, I will not contribute the money. If the organization is not successful, it will be money down the drain; if it succeeds, I will enjoy the benefits anyway. So, I will opt for a free ride (on the backs of contributors)."

This is an element of the problem that is again essentially a breakdown in the feedback mechanism. With an ordinary private good, one either pays the price and gets the good or foregoes the purchase and goes without. This is not true of collective goods. There is no assurance that paying the price will actually obtain the good for the purchaser. And there is always a chance of obtaining the good even if the price is not paid.

133

A related problem is average costs. Conservationist movements have a threshold level of time, effort, and money needed to conduct a successful campaign. If this level is not reached, expenditures count for very little; once it is reached, a momentum can be developed that may carry the campaign through to success with little additional expenditure. It follows that the more members there are to contribute money and effort, the less is the average cost to any individual member.

In a typical conservation instance, a few very dedicated individuals will spend a large amount of their resources to get the movement started. They may know that the sum of their efforts cannot possibly attain the threshold level, but they spend their resources hoping (1) that others will join them, (2) that average costs will decline, and (3) that still others will be attracted by these declining average costs. But, as observed above, the fact that the "ball is rolling" may strongly motivate free riders not to contribute. Of course, if a significant number of people attempt to become free riders, the conservation campaign is doomed. A commonplace example of this general problem is found in election campaigns. If everyone believes his candidate will win, very few will go to the polls and vote, thereby assuring the candidate's defeat. On the other hand, if everyone thinks the candidate will lose, they will not bother going to the polls to vote for him, again ensuring his defeat. If a significant number attempt a free ride, no one gets any ride at all. Yet, there are strong incentives for every individual to try for one. Therein lies the tragedy of collective goods.

LIMITED COLLECTIVE GOODS

The preceding discussion has focused mainly on *pure* collective goods — goods with zero marginal costs of use. Actually, of course, there are very few such goods. "Existence demands" for vanishing species, for great works of art, and for natural wonders account for most of them. The existence of the whooping crane, the *Mona Lisa,* and the Grand Canyon can simultaneously satisfy the demands of billions of people at no additional cost. But most collective goods are *limited.* Once their use increases beyond a certain point, congestion among users sets in, and marginal costs begin to rise. This case will be considered briefly.

    A convenient example is the overuse of a national park (Figure 3). The demand for national parks increases over time (as shown in the sequence of shifting demand curves). Up to about 3500 visitors, successive numbers of people can use the park at little or no added cost. Above this number, however, congestion begins to occur, and some means of rationing the park must be found. Thus, at D_4 in Figure 3 it is inefficient to permit more than 5000 persons to use the facility because, due to congestion, the marginal benefit of use beyond this figure is less than the marginal cost of that use. As user population expands, shown in the shifting demand curves, the optimal quantity of use also grows. The optimal use is given by the intersection of marginal cost and the relevant demand curve. But the optimum amount of use grows more slowly than user population. This is because of the rapidly increasing marginal cost curve. Eventually a point is reached (at 7000 visitors) which is the absolute limit of use. After this point, one more visitor will create more costs to others than he receives in benefits. This point is, then, the absolute carrying capacity of the park defined in economic terms.

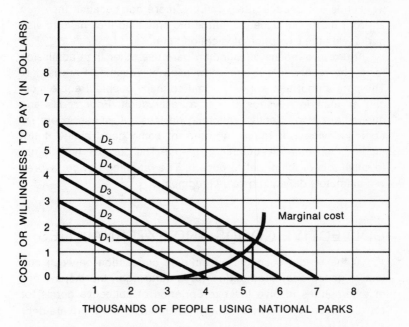

FIGURE 3: Demand and Costs of National Parks

The problem is how to limit use to the optimal amount. Many economists advocate charging an admission fee to the park in an amount defined by the intersection of marginal cost and demand. At demand D_1, the fee would be zero, but at demand D_4, the fee would be fifteen dollars. It has also been suggested that the daily admission fee should fluctuate with daily use. The fee would be very high on weekends and perhaps zero on some weekdays. This technique would presumably induce people to use the parks in a more efficient manner. Those who valued its weekend use most would be willing to pay the higher price and would use it during that period. Those who valued it less would wait for the weekdays to use it at the lower price. This changing price technique is known as "peak-load pricing."

No one denies that rationing the use of congested parks is a vital requirement. Reasonable men do, however, disagree on how to do it. The major objection to rationing by means of admission fees is that it discriminates against the poor and working people. The rich will use the park at higher fees not necessarily because they value it more, but because they can afford to pay more than the poor. Also, workingmen will have to pay more for use on weekends not because they value it more nor because they can afford it more, but because that is the only time they have to visit parks (with the exception of vacation time).

There are means of rationing park use other than admission fees. The reservation system now used at Yosemite Park is one. The park authorities will only admit so many people. People must call or mail in advance to get a reservation. Poor roads and queues are also effective rationing devices. Of course, every rationing device will ration in terms of some constraint — if not money, then time or equipment (jeeps). The purpose here is not to solve this problem but to pose it. The rationing device favored will ultimately depend on value judgments.

COLLECTIVE GOODS AND EXTERNALITIES

If the reader suspects an intimate connection between collective goods and externalities, he is quite right; the major kinds of externalities involve the expropriation of collective goods for private uses. To review, an externality results from a private decision in which the interests of some affected parties are not considered. It applies to both private and collective goods and may be either good or bad: A farmer uses a stream for irrigation purposes, and a second farmer diverts the stream for his own use without paying compensation (a private good externality); the stream has certain aesthetic and recreational qualities that can be simultaneously enjoyed by many people (a collective good); a farmer diverts this stream from recreation uses without paying compensation (a collective good externality).

In an analytic sense, collective and private good externalities are the same. They are both expropriations. However, the more serious problem of market failure due to externalities affects collective goods. While there is a pattern of incentives and laws to mitigate private good externalities, no parallel structure governs collective goods. Returning to the example of the stream, the first farmer receives substantial benefits from his use of the stream. Moreover, these benefits accrue solely to him. So when the second farmer attempts to divert the water, the first farmer is strongly induced to protect his vested interest. If the law fails to arbitrate disputes of this nature, it is not unusual for the matter to be settled on the basis of force. In Colorado, for example, there were shotgun allocations of water as late as 1965 (and perhaps later).

Contrast this with the state of the recreationist who uses the stream two or three Sunday afternoons each year. There may be a great many people like him, and the sum total of the benefits they enjoy from the stream in its natural state may greatly exceed the benefits of that water in irrigation. But no one has an interest in the stream comparable to that of the farmer. The result is that no individual enjoying the natural stream is provoked to go to great trouble and expense to protect it from diversion; by contrast, the person benefiting from its private use for irrigation has strong inducement. Even if there are a few strongly motivated Sunday fishermen who try to organize to prevent diversion of the stream, large numbers of free riders among them will prevent any effective action. In the end it is a question of whether or not feedback loops will tend to close over time. It is a question of social adaptation. Externalities are created and destroyed every day in an evolving economy. What determines their seriousness and durability is the degree to which free riders can operate.

MIXED GOODS AND THE DISTRIBUTION OF COSTS AND BENEFITS

In the discussion of the blue whale, it was noted that mixed goods can be used for either collective or private purposes but rarely for both. This set of goods defines an important domain of externalities. Severe external costs occur when a collective good is converted to private use. It is with mixed goods that this opportunity most frequently presents itself.

The provision of collective goods (or bads) is determined not by the *relation* of costs to benefits but, rather, by the *distribution* of these costs and benefits. It is quite possible that killing whales, cutting forests, and damming streams may create far more total costs than total benefits. Yet such acts will continue because the collective losses are divided among so many people that the cost to any one person is rather small. The benefits of *destroying* collective goods, however, most often accrue to only a few people, making the rewards to each quite high. Economic and political power increases as it is concentrated; it weakens when dispersed. Thus, it does not matter whether the total benefits of collective goods *exceed* their total costs. The key question is *who* receives the benefits and *who* incurs the costs. The ultimate determinant of collective goods and bads is the system of class relationships.

THE CITY AS A COLLECTIVE GOOD

In this section, the operation of this class (or power group) mechanism will be examined through reference to a central problem of our time — the progressive deterioration of cities. The contention is that there is an optimum city size (defined for each region and each set of people) but that cities tend to grow beyond this optimum when they have the opportunity. Obviously, a problem as complex as urban growth cannot be adequately treated in a few pages. But city size can serve to demonstrate the collective good roots of the problem.

Imagine a fixed amount of land onto which a progressively larger number of people are moving, along with their children, pets, houses, and the impedimenta of modern urban life. As the population grows, certain collective benefits and costs arise from the interrelationships of the new citizens. On the benefit side, there are more people to share in the indivisible costs of schools, roads, and meeting halls; more diversified school curricula can be offered; movies and the theater will appear. In the early stages of growth, each additional person represents a considerable amount of collective benefit to the whole community. In implicit recognition of this, the early settlements of the west provided strong inducements — from transportation subsidies to barn-raising parties — for new settlers.

It is also true that, after a certain degree of density is reached, additional settlers bring with them substantial costs. Their chimneys pollute the neighbors' air, their sewage pollutes the water, the noise level rises, and traffic congestion increases.

The collective benefits and the collective costs of this process can be visualized with the aid of Figure 4. The top portion

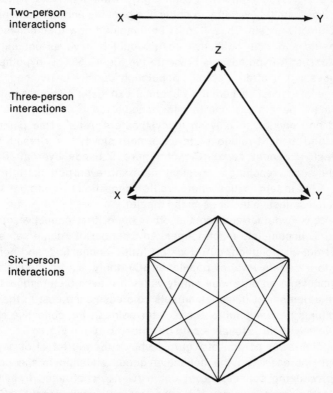

FIGURE 4: Growth in Numbers of Possible Interactions

of the figure represents a two-person settlement. Each point stands for a person (or household). Each arrowhead stands for a relationship between these people. Thus, for example, the double-headed arrow may stand for a benefit as X talks to Y and Y talks to X. Or it may stand for the cost of X polluting Y's air and vice versa. The arrows may represent either collective benefits or collective costs.

As the number of points (people) increases, however, the number of arrows (relationships) increases by a much faster rate. While in the top relationship the two persons are connected by one double-headed arrow, in the center relationship, three points are connected by three double-headed arrows; and in the lower relationship six points are connected by fifteen double-headed arrows.[3]

Mathematically, these effects will continue without limit. But, obviously, a real community cannot grow without limit, for it would become infinitely wealthy in collective benefits and infinitely poor in collective costs. Thus, it must be imagined how total collective benefits and total collective costs would behave as population increases. A hypothesis is presented in Figure 5. This hypothesis suggests that after a certain population density is reached, the range of collective benefits becomes exhausted. There are just so many people one can talk to; schools reach an optimal size; and one can go to only so many theaters. After some point is reached, benefit relations increase more slowly. As a result, the collective benefit curve (C_B) of Figure 5 bends over after an initial spurt, reaching a constant state and even perhaps turning down (as in large cities where traffic congestion discourages use of such collective facilities as the theater.)

It is hypothesized, on the other hand, that collective costs tend to accumulate much longer. Any number of people can suffer from air pollution, putrid water, traffic congestion, and noise. Up to a certain level of population (Q), the costs are negligible. Chimneys in an otherwise empty valley are even picturesque. But as the number of these "pollutants" increases, the cost to the inhabitants becomes more acute. After point Q, the collective cost curve will begin to follow a path described by C_C in Figure 5.

The lower portion of Figure 5 shows the *net* effect of population increases on the collective goods situation in the community. If the collective cost curve (C_C) is subtracted from the collective benefit curve (C_B), the net collective benefit (NCB)

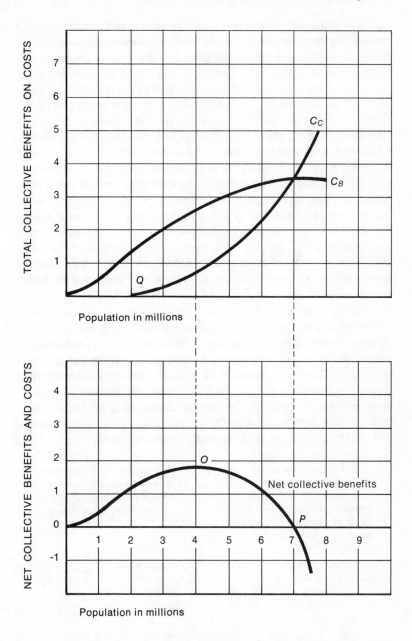

FIGURE 5: Collective Benefits and Costs Associated with Increasing City Size

curve is obtained. This curve describes an effect that seems a reasonable description of reality. As the population of a community increases, the net collective benefits also increase — at first very rapidly, then at a declining rate. Once an optimal point (*O*) is reached, however, net benefits tend to decline. Finally, the point (*P*) is reached where the collective costs of population expansion have completely offset any collective benefits that may have occurred. Further population growth will result in an absolute loss in the collective aspects of community life.

Yet, there is no reason to believe that the population growth of a city will stop at either the optimum point or even at the point where negative net benefits (net costs) begin to appear. The reason is very simple: There is still a lot of money to be made from growth. Those few who have positioned themselves so as to capture this gain will do what they can — they usually have considerable power — to make sure that growth continues. The public, lulled by the peculiar mystique of growth and burdened by free riders, pays the cost.

To say that there are inherent forces in the social and economic structure of urban dynamics which will tend to cause cities to grow beyond their optimal size is not to embrace an anti-urban philosophy. It is precisely because cities are so valuable that their tendency to destroy themselves through excessive growth is tragic. But the flight to the suburbs, the frenzied escape on weekends, and the proliferation of second homes away from cities all attest to the city's qualitative decay. The dimensions of the problem are indicated by a 1969 Gallup Poll national survey:

> . . . if a pleasant place to live were the principal consideration influencing the public, there would be a marked reversal of the trend away from rural areas. Those interviewed were shown the following list of areas and asked which of them they think would be most pleasant as places to live. Results divided [into percentages] as follows:

Rural area	30	Seashore	9
Small city	25	Large city	6
Suburbs	18	Other	1
Mountains	15		

> Analysis by size of community shows that many big city dwellers have a yearning for smaller places and those in rural areas have no yen for the "big city."[4]

MARKET FAILURE AND FREEDOM

In cases where the results of one person's decision are determined by what other people do, it may be impossible for an individual to make a rational decision without a coordinating social rule. As in the instance of the blue whale, individuals may be induced to hide their real preferences, either in hope of becoming free riders or from fear of being victimized by free riders. An individual may then rationally decide to curb his own freedom of choice in order to establish a social rule by means of which he can act more reasonably. For example, one may rationally refuse to donate $20 per year toward the preservation of the blue whale and then vote for a tax on himself toward this end. The tax prohibits free riders. The purchaser's chances of getting what he paid for — the preservation of the whale — are likely to be much greater with a tax than by private donations. It is also likely that with a tax he will pay much less to "purchase" his share of the collective good. Since everybody will have to pay, the cost per person is lower. Of course, taxation always carries a certain inefficiency in that some people who do not value the particular good have to pay anyway. But this inefficiency is alleviated, though certainly not eliminated, in the political process of exchange: "You vote for my blue whale, I'll vote for your moon shot."

Similarly, an individual fisherman who kills every whale he can may also rationally vote in favor of laws regulating the number he can kill. It is obvious that if the whale becomes extinct, so will his profession and livelihood. But it is equally obvious that if he foregoes a kill in the long-run interest of his business, a free-riding fisherman will likely come along and kill the whale that was supposed to be saved. Every individual fisherman may, however, rationally vote to limit his kill if he is assured everyone else must do the same.

Thus, a final irony is encountered in the perplexing subject of market failure. It is the *paradox of freedom;* If everyone is free to do as he wishes, no one will be free to do as he wishes! But this paradox, like most others, rests on a fundamental misconception of the problem and a misuse of words.

If freedom is defined as the "absence of constraint," the paradox of freedom inevitably follows. But if freedom refers to the range of choice open to all individuals in a society, it is easily seen that freedom can be increased by curbing certain choices. An elementary example is the law requiring everyone to drive on the same side of the street. The law does not allow the freedom to choose either side. But it assures that one has a much wider range of choice — that the freedom to travel safely in an automobile is maintained. There are many such instances where the acts of individuals are so interdependent that no single person can make a rational choice without some assurance regarding the behavior of others.

Adam Smith thought that each man pursuing his own self-interest is led as though by an unseen hand to promote the interests of all. But science is beginning to discover many different unseen hands in economic and environmental structures. Sometimes these unseen hands lead the pursuit of self-interest not only against the interests of others, but even against the interest of the choosing individual himself. For this reason, people are beginning to realize that it is only through the judicious control of certain choices that the ideal of maximum choice can be reached.

FOOTNOTES

[1]Collective goods are also called public goods. We use the former term here in order not to confuse this specific kind of market failure with other kinds of publicly provided goods. For basic literature on collective goods, see W. J. Baumol, *Op. cit.*; Mancur Olsen Jr., *The Logic of Collective Action* (Cambridge: Harvard University Press, 1965); James M. Buchanan and Gordon Tullock, *The Calculus of Consent: Logical Foundations of Constitutional Democracy* (Ann Arbor: University of Michigan Press, 1962).

[2]There is every reason to believe that these demand curves slope downward to the right in the normal way, reflecting diminishing values of successive warnings. It could be argued that the demand curves are not this way but are horizontal, reflecting a constant value for each use. This would be a complex debate irrelevant to the present purposes since it makes no basic difference to the ultimate analysis.

[3]Generally, the relation of arrows (*A*) to points (*P*) is given by the formula

$$A = \frac{P(P-1)}{2}$$

[4]U.S. Congress, House, Committees on Government Operations, Study for the National Wildlife Federation by the Gallup Poll, *The Environmental Decade, Hearings* (Washington: U.S. Govt. Print. Off., 1970), 91st Cong., 2nd sess., February 2, 3, 4, 5, 6; March 13 and April 3, 1970, p. 21. (The numbers add to 104 in the original.) Of course, these results are indicative only. Urban dwellers receive higher incomes which may compensate them for living, by their own testament, in less desirable cities. It would have been interesting to see how much urban people would be willing to sacrifice in income to live elsewhere.

chapter ten

The Logic Of Conservation

INTRODUCTION

So far the discussion has concerned problems that occur (or are assumed to occur) in a single time period. The length of the time period can be an instant, a year, or even a century. But in each case problem and cure were confined to the single period. In economics, such analysis is called *static analysis.* In environmental economics, events that are causally related through more than one period of time must be considered. This multiperiod study is called *dynamic analysis.* The transition from statics to dynamics involves more than adding together individual static time periods. The introduction of time into economic analysis demands rather radical change of disposition.

"Dynamics" is, unfortunately, not a settled term and can be interpreted in at least two ways. It can mean the study of a given system as that system grows over time. In this case, the basic features of the system are known; all that remains is the determination of how the system will work out its adjustment problems, or, in terms used earlier, how it will move toward equilibrium. Dynamics can also mean the study of the time path of a system whose structure is changing over time. In this case changes may be occurring in tastes, technology, or even in the form of economic organization. It is easy to see that this second kind of dynamic analysis poses more formidable difficulties than the first. The first is mainly a problem of determining logical mathematical and engineering sequences flowing out of a given and well-understood system. The second entails all of this plus the problem of predicting (1) when changes will occur in the system, (2) how severe the changes will be, and (3) what kinds of changes might occur. One has only to contemplate the problem of predicting changes in tastes, technology, and institutions to understand what a very unsettling business this can be. In order to develop rational plans for the future environment, an attempt at dynamic analysis of this second type must be made. It is known, as a matter of historical record, that the system does change. Any plan that does not explicitly acknowledge this is almost certain to be incomplete, if not totally erroneous. Planning the future environment is an attempt to make systematic decisions under conditions of uncertainty or ignorance. This central problem underlies most conservationist programs and provides the central theme for this chapter.

THE BLUE WHALE REVISITED

The blue whale will again be used as an example of conservationists' problems. Imagine the time to be 1920 and that the members of the whaling industry have as their objective catching all the whales they can to supply the market for whale oil. But a dreamer among them points out that, if the current (1920) rates of whale kill continue, the species will be driven to extinction in a matter of a few short decades. Being well trained in the practice of economic rationality, the industry's economists quickly compute the present value of this resource. They find that, at a discount rate of six percent, the 1920 value of one dollar in 1970 is only five cents. Thus, a current 1920 whale is worth roughly 20 times what a 1970 whale would be worth. Now the question is asked: Is it likely that whales will rise in value twentyfold over the next 50 years? Hardly! Even in 1920 electricity and petroleum are beginning to compete with whale oil. The whalers, if they are rational, had better get all they can while they can. Exit the dreamer accompanied by derisive noises from the practical-men-of-affairs in the chorus, who thereupon proceed to exterminate whales with customary fervor.

It is not easy to fault these early whalers for their treatment of this problem of future-versus-present whales. Their discounting procedure is immaculate. It is exactly the way in which decisions are made regarding the allocation of resources to bridges and dams today. Yet if these early whalers had been equipped with 1970 whaling technology and sufficient demand for whale oil, this form of economic reasoning would have led them very quickly to hunt the whale to extinction. The few whales that do remain owe their existence more to technological incapacity than to any rational conservation policy among whalers or governments.

This simple example demonstrates the inadequacy of the decision-making apparatus. Something is clearly amiss when the relations between present and future are decided using only an elementary bit of discounting arithmetic. Must the whaler's argument be accepted as complete? The answer is, of course, no. But in order to arrive at this answer, a major assumption of much of the economic analysis used in this book must be relaxed — that is, "perfect knowledge." When this assumption is eased however, many complications follow.

If the present and future values of whales (whether for oil, meat, or as collective goods) are known with certainty, the correct value of whales at any point in time can be calculated using the appropriate rate of discount. Moreover, if it is then found that whales should be used to the point of extinction, this conclusion would have to be accepted as a correct application of economic reasoning. If the conclusion is not accepted, the rejection must be based on other grounds. One could argue that it is immoral to exterminate a form of life. But except for this (valid) kind of appeal, the economic conclusion would be essentially correct.

But man does not have perfect knowledge. He knows very little about the present and much less about the future. Since little is known about the future, it is virtually certain that man will make mistakes with respect to it. Society will inevitably regret some of today's decisions. There are two kinds of mistakes one can make. The distinction between them is critical to a discussion of conservation. One can make a *reversible* mistake where once the mistake is found out, one can with some time and perhaps trouble correct it. But one can also make *irreversible* mistakes from which there can be no recovery. It need hardly be pointed out that if mistakes must be made, the former are very much preferable to the latter.

The early whalers' decision-making process went awry in not considering that their actions might be irreversible. They assumed only the traditional private good market for whale products. Moreover, they thought they knew enough about the future to correctly decide to capture more whales. They failed to distinguish among use, preservation, and extinction. In a world of certainty, these terms could be comparable. In the real world of uncertainty, they are basically noncomparable. The alternatives differ demonstrably with respect to their degree of reversibility. The case of the blue whale shows this quite clearly. As long as the whale population is above the minimum threshold at which the whales can reproduce themselves, mistaken policies with respect to use and preservation are reversible. If too many whales are captured in one time period, whaling can be decreased in the next time period, and the population will reestablish itself. But whaling to such an extent that the population extinction threshold is approached introduces the element of irreversibility. Once this threshold is crossed, mistakes can never be rectified.

In deciding when to use its valuable resources, society is often pulled in opposite directions. Because more certainty surrounds the present, there is a strong preference to use resources now rather than save them for the clouded future. This causes no major problems for society as long as it is engaged in reversible courses of action. As activities become less certain and as the possibility of irreversible consequences increases, society must take pains to keep open the option of future use. The great bulk of conservation controversies have revolved around this very issue.

THE HELL'S CANYON CONTROVERSY

A contemporary instance of the present-versus-future conflict in conservation arises in connection with the use of natural valleys for reservoirs to meet the needs of irrigated agriculture, navigation, and the production of hydro-electric power. John R. Krutilla of Resources for the Future has recently been involved in an important study of such a problem related to the development of the Hell's Canyon region of Idaho.[1] Krutilla's analysis is presented here in some detail because of its applicability to a wide range of similar problems.

Hell's Canyon is popularly valued in its natural state as a source of scientific investigation, aesthetic appreciation, and recreation. But it also has a high commercial value as dam sites for electrical production. The demands for both the "natural" and "commercial" uses are expected to grow. Unfortunately, there is only one Hell's Canyon. If the area is to be used as a one-of-a-kind natural site, it cannot simultaneously be used for power production or irrigation or other commercial use. While the natural use cannot be duplicated or supplied elsewhere, there are many substitutes for the commercial use of the canyon. Electricity, for example, can be generated from fossil fuels or nuclear energy. Moreover, as technology expands, substitute sources of energy should reduce the value of Hell's Canyon for commercial purposes.

An economic analysis of this situation is presented in Figures 1 and 2. Each supply and demand curve in Figure 1 is dated from the present (period t) to the next year ($t + 1$) to the year after that ($t + 2$). Clearly, the demand for electricity expands over future years, but the supply expands even more rapidly as technology develops. Since supply is outrunning demand, the price

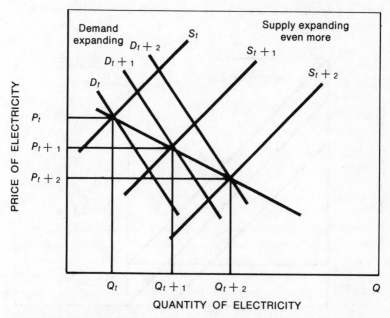

FIGURE 1: Hypothetical Price Trends of Commercial Use of Hell's Canyon

or relative value of electricity falls over future time periods ($P_t > P_{t+1} > P_{t+2}$ in Figure 1).

In Figure 2, however, quite a different situation is encountered. Here demand is also rising through time, but the supply of the natural use of Hell's Canyon is fixed and cannot be expanded by technology. As time goes on, more and more people enjoy Hell's Canyon in its natural state; and this increasing demand works against the fixed supply curve. This leads to a continual increase in the value of the canyon as a natural area.

The relative values of the two uses of this resource diverge over time. The value of Hell's Canyon as a natural area increases each year while its value in commercial uses falls. In order to arrive at the optimal use of Hell's Canyon, estimates of the relative rates of increase and decrease in value for each potential use must be made. Moreover, these streams of values through many periods must be discounted to find a present value of each type of long-term use. No one really knows what these values are. But Krutilla has devised an ingenious approach to this problem. In effect, he says, it may not be necessary to know with precision what the *total* values in each use are; it may be sufficient only to know that the *current* value as a natural site exceeds a certain minimum.

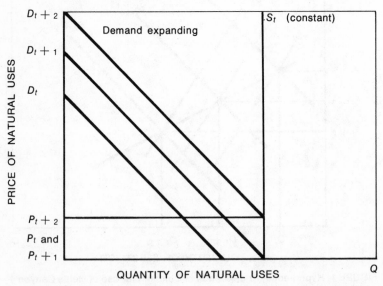

FIGURE 2: Hypothetical Price Trends of Natural Use of Hell's Canyon

Krutilla approaches the question in this way: (1) through the ordinary techniques of benefit/cost analysis, estimates of the present discounted value of the commercial use of Hell's Canyon can be made; (2) by estimating future population growth, increased incomes, increased leisure, and other relevant variables, the increased demand for the natural use of Hell's Canyon in the future can be estimated; (3) given these estimates of increasing demand for its use as a natural area, the rate of appreciation in value of Hell's Canyon can be determined; (4) then it can be asked, given this rate of appreciation in value, for natural uses, what must the minimum current value be in order to make the present value of the natural use of Hell's Canyon equal to or greater than the present value of its commercial use?

Such a procedure is based on assumptions about the rate of discount, the rate of technological advance in electrical power production, the rate of increase in recreational uses of natural areas, and the time span that is to be considered. Krutilla's estimates of these components vary, but consistantly show that the use of Hell's Canyon as a natural area has a sufficiently high value to offset values created by developing the area for commercial purposes. That is, the demand curve for natural uses will move up the vertical supply curve rapidly enough to provide more consumer's willingness to pay than would be generated by the continually changing demand *and supply* curves for power. The relative changes are sufficient to rationally decide that the best use of Hell's Canyon is as a natural area for recreation and scientific purposes.

CONSERVATION

The preceding example shows how single-purpose economic and political systems can fail in the optimal allocation of resources over time. The plight of the whale and the whalers reinforces the conviction that the unintended consequences of rational choice must be studied. Krutilla's analysis shows that the current value of the natural use of Hell's Canyon far exceeds the commercial value of that site, yet it was a very close contest between conservationists and the power interests that finally resulted in the preservation of this resource in its natural state. It is not difficult to see why "the system" could fail in either of these cases.

All the benefits of power development accrue over a rather short period of time. The benefits of natural use accrue over a much longer period of time and include successive generations of people among its beneficiaries. In order for the market to perform well in this situation, it would be necessary for all the future beneficiaries to engage in the bargaining process. It is not enough for merely the present population to be willing to pay for the natural amenities of Hell's Canyon, or to cast votes in its favor; all future beneficiaries must also do so. This is obviously impossible, yet for proper allocations between present and future, the interests of the unborn generations must be taken into account. This is what Krutilla's economic analysis attempts to do.

Unfortunately, resources are as often allocated in terms of political and economic power as in terms of economic efficiency. The brutal fact is that future generations have neither money nor votes. This fact has led many thoughtful people to declare it the state's responsibility to look out for its future citizens — if necessary, against the wishes of its present citizens.

OPTION DEMANDS

The neglect of future generations is one reason why the market fails in allocations between commercial and natural uses of mixed resources. Another reason is that the allocation of resources over time often involves a very special kind of collective good.

Imagine that you never expect to visit Hell's Canyon, yet this natural amenity is threatened with extinction by hydro-electric development. Now even if you do not expect to see Hell's Canyon, you may be willing to pay a defineable sum to see it preserved. It may have an existence value to you just as the blue whale has existence value to Joan Baez. But it may also have a very important *option value* — although you may not expect to see it, you may want to preserve the "option" to see it if present conditions change. Or you may be willing to pay some amount of money to preserve the option for your children or their children (or even unrelated future generations) to see it.

Such option demands are collective goods. A large number of people can enjoy the option simultaneously. Thus, the sum of the option demands for natural uses of resources, such as Hell's Canyon, can amount to a very large figure. Yet, for reasons explained above — namely, the free rider — few, if any, of these option demands will become truly *effective* demands.

Finally, it should be noted that the essential reason for maintaining an active option demand is the threat of *irreversibilities*. One would not be willing to pay for an option if the process were reversible since one would already have the option. Thus, while there are option demands to preserve Hell's Canyon in its natural state, there are no option demands to "keep it" for hydro-electric production. If kept in a natural state, the canyon can always be developed for power production in later years. The converse is not true. Once the reservoir fills, the natural amenities of the resource are lost forever. For this reason the present value of the commercial use of Hell's Canyon must exceed its natural value; one would want a large premium over the natural value to embark on such an irreversible course of action as building a dam.

CONCLUSION

Since very little is known about the future, it is wise to choose reversible actions whenever possible. The demands of conservationists to keep unique features of the environment in their natural state is consistent with this criterion. If the redwoods are "saved" and society later finds this to have been a mistaken decision, the trees can always be made into lumber. Redwood fence-posts cannot be made back into trees. The analysis of Hell's Canyon showed that the direction of change of relative values can favor the natural state of unique environmental phenomena rather than their commercial uses.

Man's inevitable ignorance about the future and his poor knowledge of what is reversible and what is not suggest that considerations of economic efficiency work toward preservation of the unique features of the natural environment. This is but another way of keeping open a wide range of choices and maintaining for man the varied kind of environment that he now enjoys.

FOOTNOTES

[1]Krutilla's research has not yet been published. Various aspects of it and alternative formulations of the dollar requirements have appeared in papers delivered to classes, conferences, and seminars. It is likely that a systematic treatment of his efforts will soon be published by Resources for the Future, Washington, D.C.

part three

chapter eleven

Recreation And Cost/ Effectiveness: Some Further Applications

The last three chapters have been devoted to serious economic problems that are a part of the contemporary struggle with environmental deterioration. For the most part the problems have been analyzed in market terms. Even the consideration of externalities was eventually resolved through market or market-like mechanisms. This chapter concentrates on extensions of nonmarket analyses that may be useful in coming to grips with environmental problems. The chapter opens with a brief set of definitions of nonmarket goals and phenomena then turns to economic analysis as it has been conducted in the area of outdoor recreation. Results of such research provide imputed "prices" even though no true markets exist. Prices derived from this analysis can be used for benefit/

cost determinations in public policy decisions or they can be used in an even more difficult decision-making context: cost/effectiveness. The final section of the chapter discusses the relationship between the distribution of income and public decision-making.

Some classes of goods and services are virtually impossible to evaluate even through such conventional nonmarket devices as benefit/cost analysis. Goods and services of this type can be *intangible* or *incommensurable.* The difference between the two is considerable. An intangible is defined as something that cannot be measured in quantitative terms. Incommensurables come in sets of two or more and are so-called because the items in the set cannot be measured or compared using the same standard. Color is an intangible when placed in a context of beauty or artistic composition. It simply has no quantitative measure. Similarly, the spectacular view of a deciduous forest in the fall months has no quantitative measure and is intangible. A brick building provides a pair of incommensurable measures. The bricks in the building can be measured in terms of their weight and this weight can be translated into quantitative physical forces acting on the ground that must support the building. Or the bricks can be measured with reference to their capacity to earn rents and profits for the man who stacked them in a certain way to create office space. The two quantitative measures — weight and rent — derive from the bricks, but they cannot be compared; they are incommensurable. Using an example more familiar in economics, a laborer may through his life's work produce services valued at $500,000; but it would be absurd to say the laborer's life is *worth* this amount. The two things cannot be compared. They, too, are incommensurable. This chapter deals with the application of economic and "para-economic" techniques to situations in which intangible and/or incommensurable items are involved.

As governments have turned their attention to controlling the deterioration of the environment, they have been forced to look for analytic and scientific devices that permit discussion of intangible and incommensurable items. Early efforts at finding surrogate measures of intangible values came in response to the need to have a measure of benefits for benefit/cost analyses (see Chapter 7). Among economists who attempted to fill this requirement were a number working in the area of outdoor recreation. Their efforts will be examined here. Outdoor recreation is not being presented as the best vehicle for a discussion of intangibles and incommensurables; it is used because of its accessibility in the literature and its closeness to the environmental quality theme.

Like beginning work in most fields, the efforts in outdoor recreation so far have been somewhat crude. Appropriately enough, the National Park Service of the United States Department of the Interior initiated the efforts to determine the dollar benefits coming from recreation facilities. On June 11, 1947, A. E. Demarey, then Associate Director of the National Park Service, wrote to ten professional economists with competence or interest in recreation.[1] His letter suggested that the National Park Service was interested in conducting a comprehensive economic study of the National Park System, but apparently the "comprehensive" study was to be limited to answering these questions:

1. What is the dollar volume of secondary or indirect benefits[2] produced by National Parks, and by how much does this exceed the economic returns and benefits that would accrue if the land and resources in the parks were used for other purposes?

2. What is the dollar value of the National Parks to the nation, states, and local areas?

3. What is the volume of economic development occurring adjacent to National Parks, and is there a relationship between the development of parks and the intensity of economic activity in nearby areas?

Nine of the ten economists responded to the inquiry. Two respondents gave no specific answers to the questions as posed; five suggested that surveys be taken to determine consumer expenditures; and two people, Professor Howard Ellis of the University of California and Professor Harold Hotelling of the University of North Carolina, made positive suggestions about the application of economic reasoning to the questions at hand.

Both Ellis and Hotelling interpreted the questions to mean that the National Park Service was seeking a monetary value for the park or for the experience the park offers. While this monetary value is clearly a value on an intangible item, it could be usefully converted into an estimate of benefits for a benefit/cost ratio and could be of considerable help in decision-making processes. Ellis suggested that the only effective way to find such a value would be to charge an entrance fee and allow each individual park to be operated as though it were a private business with profits going to the government.

Hotelling, on the other hand, suggested locating the recreation facility on a map and drawing concentric rings or zones around it so that everyone residing in a particular zone would incur approximately the same travel costs if he were to visit the park facility. Those living in the more distant zones would pay high prices, while those living nearby would obtain the visit at bargain prices. Hotelling reasoned that by carefully interpreting the cost, distance, and population variables, a suitable demand curve could be developed.

Table 1
Hypothetical Data on the Demand for Recreational Services Provided by a Particular Site

Zone	Population	Visitors to park	Visitors per 1000 population	Expenses incurred per visit
A (100 miles distant)	3000	15	50	$20
B (200 miles distant)	4000	16	40	25
C (300 miles distant)	5000	15	30	35
D (400 miles distant)	6000	12	20	40
E (500 miles distant)	15,000	15	10	55

About a decade after the National Park Service made its survey, a number of scholars attempted to use the Hotelling approach to determine demand curves for recreational facilities. The most popular attempts were made by Marion Clawson.[3] Clawson was searching for a way to use existing data to develop a hypothetical demand curve for a whole recreation experience (such as a trip to several parks) and simultaneously to develop a hypothetical demand curve for a single park or facility. To do this, Clawson classified the visitors to a park into zones of origin and calculated total expenses per trip for persons coming from each zone. The unique part of the Clawson approach is that he placed all visits in standard terms by relating visitors from a zone to population in that zone — visitors per thousand population. Hypothetical data for all items mentioned above are presented in Table 1. As distance from the park increases, the cost of a visit increases. Even though there is not great variation in the number of total visitors coming from each zone, the visitors per 1000 population decrease as expenses per visit increase. This latter relationship provides the basis for Clawson's first demand curve — the one relating to the experience as a whole. Such a curve is presented in Figure 1. The solid line in the figure reflects the data in Table 1. The broken line is a summary curve used to "smooth out" the first approximations of the demand curve.

If all other factors remain constant, most groups of people will respond to price changes in about the same way. Using this basic assumption, Clawson derives a demand curve for a particular site from the demand curve for the whole experience. The derivation begins by attaching an imaginary fee to the site in question, then inquiring as to how populations in each of the zones would respond to such a fee. Anticipated responses are determined by reference to the broken line demand curve in Figure 1. It can be seen from this curve that, if the total expenses required to visit a park are $20 per visit, visitation will occur at the rate of 45 persons per 1000 population. If the price were to increase to $25 per visit as a result of a fee being imposed, vis-

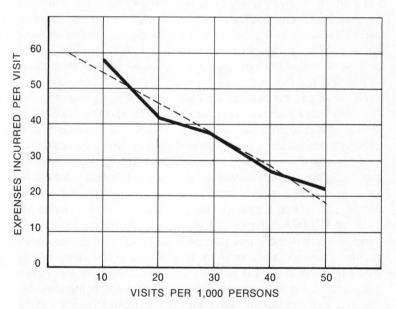

FIGURE 1: Demand Relationship Derived from Hypothetical Data in Table 1

itation rates would drop to 40 persons per 1000 population. The response of the hypothetical populations of Table 1 to a five dollar fee is indicated in Table 2. The new visitation rate is developed from the demand curve in Figure 1. To obtain this new rate, the fee imposed is regarded as a normal expense incurred in visiting an area. The resulting visitation rate can be multiplied by the population to yield the total park visits that would be made from each zone if a five dollar fee were charged for the use of the facility. The visits from each zone can be summed to obtain total park attendance. In the hypothetical case above, total attendance at zero fee would be 730 (Table 1), while total visits with a five dollar fee would be 525 (Table 2).

These two price and quantity relationships may be considered points on a demand curve for a particular recreational facility. The zero price attendance of 730 persons reflects that point on the demand curve at which the curve intersects the horizontal axis. At this point, the marginal utility of an additional visit is zero, so the last visitor received utility valued at zero and would not have visited the park if a price greater than zero had been charged. Another known point on the demand curve indicates that 525 visitors received utility of at least five dollars. By using this process and by imposing other levels of fees on the demand curves depicted in Figure 1, more points on the second demand curve can be established. Ultimately, enough of these points emerge to allow the approximation of a second demand curve — the one for the individual recreation facility. This newly derived curve can be used in the same way as any demand curve. The significant point is, however, that using existing information Marion Clawson has put monetary "values" on a good that is generally described as intangible and even incommensurable.

There are weaknesses to this method. The costs of gasoline, tires, lodging, and restaurant meals are used to find the demand for a park that provides scenic beauty, relaxation, and recreation. It also yields a peculiar logic in that if no travel costs are incurred to reach a park — as with most neighborhood and city parks — no demand curve can be developed. Still the approach deserves mention since it gives an example of the kinds of analytic exercises that must be developed if intangibles and incommensurables are to be considered by planners and politicians.

The demand curves developed for nonmarket items can be used to predict price - quantity relationships, responses to changes in price, or total willingness to pay. The last item is found by examining a demand curve and summing the amounts each individual would be willing to pay for an item rather than go without. It is the value of the total triangular area under a demand curve (such as the broken line in Figure 1). This total willingness to pay can be used as an estimate of benefits in a benefit/cost analysis. Hence, many items that ordinarily escape economic analysis because they lack ordinary market prices can be examined.

Table 2
Derivations of Demand Curve for Individual Site

Zone	From Table 1	
	Visitors per 1000 population	Expenses incurred per visit
A (100 miles distant)	50	20
B (200 miles distant)	40	25
C (300 miles distant)	30	35
D (400 miles distant)	20	40
E (500 miles distant)	10	55

After imposition of a $5.00 fee			Number of visitors with fee
Expenses incurred per visit	Visitors per 1000 population	Total population	
25	40	3000	120
30	35	4000	140
40	20	5000	100
45	15	6000	90
60	5	15,000	75

Many economists contend that benefit/cost analysis can lead to correct evaluations in some cases. But in others, the burden of intangibility and incommensurability is too great. In such cases benefit/cost analysis is a necessary, but rarely sufficient, ingredient of evaluation or decision making.

Especially serious problems arise in attempts to evaluate or compare investment activities (public or private) having more than one objective. A park may be developed for scientific research (an arboretum) as well as for recreation. Similarly, it is very difficult to compare a park planned primarily for use by small children with an equally costly facility planned for the elderly, even when measures of willingness to pay are available for each use. The lesson is that though partial solutions have been offered for the intangibility problem, it still confounds decision-makers. The use of benefit/cost analysis has been expanded into a tool called cost/effectiveness to aid in the resolution of such quandaries.

Cost/effectiveness was pioneered by the Department of Defense in the 1950's and early 1960's. The logic of the tool is simple even if the practice is not. It merely says that if the decision maker limits his considerations to costs, then he must disregard effectiveness. This is a common phenomenon. For example, Congress may allocate funds for a program designed to alleviate hunger in, say, Appalachia. The funds represent only costs; the effectiveness of the program is disregarded. One and one-half times as much money might have tripled the effectiveness of the program, but, since only cost is considered, a poor decision may have been made. If only effectiveness is to be considered in the decision, then costs must be disregarded. If the United States wants a complete and fail-safe program of defense against nuclear attack, for instance, cost must be disregarded; only the effectiveness of the system is at stake. Cost/effectiveness was designed to simultaneously examine two sides of a single coin. It is a study of trade-offs or opportunity costs. While the concept of opportunity costs dates to the early years of economic theory and several alternative formulations of the concept exist, each has at its core the idea that in deciding to produce one thing, the opportunity to produce something else disappears. The cost of producing apples can be described as so many pears given up. More recently, the concept has been popularized through use of the term "trade-off." In a study of the environment, this notion has real appeal. An analytic scheme that includes trade-offs is superior to one that does not, and cost/effectiveness appears to be an operational first approximation of such a scheme.

The first significant use of cost/effectiveness in government decision making was probably in the Department of Defense when that department was under the leadership of Robert Mc-Namara. When he became Secretary of Defense, McNamara attempted to end the century-long trap in which overall defense plans were made by building up from individual subdepartment requests for things like guns, tanks, destroyers, and bombers. In effect, McNamara asked for a unified plan exhibiting a certain degree of defense capability (effectiveness). More than this, the costs of altering this plan were to be specified in capacity gained or sacrificed as well as in dollar terms. McNamara's planners had little past experience but used their knowledge of opportunity costs to construct trade-off curves that became the first large-scale application of cost/effectiveness. After the cost/effectiveness idea had begun to take shape, it was clear that multiple goals could be simultaneously considered. Defense was the prime goal of the Department, but subgoals such as capability with nuclear submarines, size of infantry, and missile striking force were essential before the full system could be used.[4] The notions lying behind cost-effectiveness can be illustrated in terms of one of the more difficult problems of environmental (or any other) economics — the conflict between "willingness to pay" and the distribution of income.

INCOME DISTRIBUTION

Modern economics makes a very firm distinction between the economics of efficiency and the question of income distribution. How to produce goods and services most efficiently is one question. Who should get the income generated by the production of products is another. Many economists believe that the distribution of income in a society is a subject beyond their range of expertise. But it is essentially the same question as the distribution of goods and services, since income determines how much and what a person will buy. It is ultimately a value judgment.

Yet, this value judgment goes a long way in explaining who will get the output of society. It is easy to see that efficiency and distribution are *not* independent items. In fact, there are many efficient positions for an economy, one corresponding to each distribution of income. Under one distribution of income, the tastes of one class of people will predominate in the market, and under another distribution other peoples' tastes will predominate. These differing tastes will be reflected in different demands, different market prices, and different collections of goods. The result is that a variety of production patterns, corresponding to different distributions of income, can be considered socially optimal. It depends on what is believed to be most valuable. Such values, of course, cannot be determined by an economist. He must be told what is desirable by a social decision maker — the President, the Congress, or someone else. Often the disposition of decision-making groups about what is best for a society is ambiguous. No one really knows what is the most desirable distribution of income; hence, no one knows what is the best collection of goods and services.

The problem of distribution becomes critical in the evaluation of publicly provided recreation facilities or any other intangible or incommensurable product. An example is the municipal zoo in Denver, Colorado. The Denver Zoo is located within comfortable walking distance of some of the poorer areas of the city and for many years was a popular recreational facility for the residents of these areas. No statistics are available on the incomes of users of the zoo, but many of them are obviously rather poor. A good portion of the users are nonwhite — Black and Chicano.

Prior to 1966 the Denver Zoo was free. In June, 1966, an admission fee of 50 cents was imposed for everyone over 16 years of age. As an economist would expect, initiation of an admission fee caused the numbers of visitors to drop.

June through January	Total Attendance
1965-66 (free admission)	840,513
1966-67 (50 cents admission)	397,101
*1967-68 (50 cents admission)	404,149

*January estimated

The 50-cent admission charge cut attendance by somewhat more than 50 percent. And it was not simply a temporary reduction in use (as many administrators thought it would be), but the reduction continued through the second year of fee charges. One of the authors observed the rationing effect of this fee. On one occasion the older members of several families (mainly Black) remained outside the tollgate while their children (those apparently under 16 years of age) visited the zoo. The admission fee, in effect, created an income rationing device on the zoo's visitors. In their imposition of a fee, the Denver city and zoo officials may have met one objective — that of increasing revenue — but they surely detracted from other objectives. This trade-off between objectives is not ordinarily reflected in decision-making tools.

Benefit/cost evaluation of the Denver Zoo may have yielded sufficient information to convince administrators that the fee was warranted or even desirable. Many advocates of the benefit/cost evaluation of outdoor recreation, however, contend that the tool should be used only for evaluation and not to answer the question of whether or not fees should or can be imposed. There is an element of truth in this, but there is also an element of error. If investments in outdoor recreation facilities are determined on the basis of willingness (ability) to pay, then (other factors equal) the selection of investments will tend to favor those facilities used by wealthy people over those used by poor people. This is true simply because the rich are willing to pay more and, hence, imputed benefits are higher.

Professors Herbert Stoevener and William Brown of Oregon State University have conducted an interesting experiment showing the effects of income distribution on evaluation.[5] They used as their measure of benefit the imputed "willingness to pay" of anglers fishing the rivers of Oregon for steelhead and salmon. After finding that the average annual income of the fishermen was $6954, benefits for the average income and for groups of fishermen assumed to have incomes higher and lower than this average were computed. The results are shown here:

Assumed average income of users	Total consumers' surplus accruing to the fishery
$ 4000	$3,770,000
$ 6954	$5,770,000
$10,000	$7,770,000

If users of this fishery had an average income of only $4000, the imputed benefits would be less than one-half the benefits of the *same facility* serving the *same people* if their average incomes were to increase to $10,000.

The purpose of these examples is not to prove that outdoor recreational facilities should be designed for either the rich or the poor. Rather, they show that there is no *objective* way to evaluate and compare such activities and facilities. It is ultimately a question for the public, hopefully represented by competent public officials, to decide.

Economic evaluations, whether made for profit motives in the private sector or for the public good by elected officials, have two sides. These are benefits gained and costs paid. An important part of the economist's job is to identify the relationship between what is being received and what is being paid. One way to do this is to use a kind of cost/effectiveness procedure and define a trade-off function between various objectives. A trade-off function in which the two objectives are revenue from the collection of fees and visits by tourists is shown in Figure 2. The vertical axis shows increasing revenues earned by charging entry fees to an outdoor recreation facility. The horizontal axis meas-

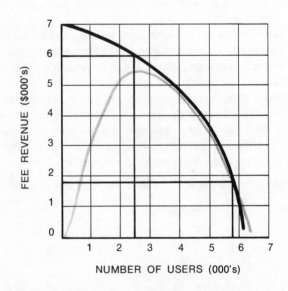

FIGURE 2: Trade-offs Between Fee Revenue and Number of Users

ures the number of people who will use the facility at different fees. At a high fee, only a few people will use the facility, and revenue will be low. At a low fee, a great many will use it, including many poor people who were filtered out by the higher fee structure, but revenue will be low. No economist can say which is the "correct" fee to charge or which is the optimum trade-off between visitation and revenue.6 The important thing is the increase in information available to decision makers.

The principle can be applied to other kinds of environmental problems. Economists are often asked to determine the optimal level of pollution control. As demonstrated in Chapter 8, an economist can establish a trade-off function between the production of goods and services and varying levels of pollution control, but he cannot say what the correct or best position on that trade-off curve is. Pollution causes a closing of options, such as loss of life and health, that cannot readily be measured in dollars. Trade-offs must be determined by the public in light of its subjective values and ethics. A portion of these values may be accurately reflected in willingness to pay. But to repeat the fundamental point, willingness to pay is never a *sufficient* basis for decisions regarding public affairs.

A recapitulation of the essential features of the Hell's Canyon study (Chapter 10) now may be used to further illustrate the usefulness of trade-off curves similar to those proposed above. The trade will be between Hell's Canyon as a commercial power generating facility and as a natural recreation and scientific area. The commercial and natural uses of Hell's Canyon have a common objective — to contribute to the economic welfare of the nation as that welfare is reflected by willingness to pay for electricity and willingness to pay for recreation. Willingness to pay constitutes one *standard of value* by which the two potential uses of Hell's Canyon may be compared.

The commercial development of Hell's Canyon for power production is a *single objective* alternative — to provide electricity (and perhaps an incidental amount of reservoir-based recreation) properly evaluated by the public's willingness to pay for this product. The natural use of Hell's Canyon is multipurpose, but it, too, creates "market values," measured by the public's willingness to pay for recreation, for scientific research, for a sanctuary for wildlife, and for option and existence values remaining for present and future generations. Even if one believes that the noncommercial objectives (laboratory and sanctuary) could in principle be evaluated by some willingness to pay scheme, it is clear that no actual measurement is presently possible. These objectives must be *independently evaluated* in terms other than willingness to pay. In effect, the alternatives reduce to this: A single-purpose commercial use is to be compared to a multipurpose natural use.

If the market value of the natural use is greater than or equal to the market value of the commercial use, then that is a sufficient basis to choose the natural use. When the natural use is selected, the willingness to pay on the part of recreationists carries with it the use of the area as laboratory and sanctuary. If, however, the case were reversed and it were found that the market value of the commercial use exceeded that of the natural use, that is not sufficient to decide in favor of the commercial use. The value of the nonmarket objectives served by the natural use may, when added to its market value, exceed that of the commercial use. Since these nonmarket values are not objectively known, the ultimate decision must rest with the essentially subjective evaluation of a decision maker.

In sum, an economist's attempt to determine the market value of outdoor recreation (and other features of the environment) through estimates of willingness to pay is an essential feature of the decision-making process. But it is never a sufficient basis on which to make the ultimate decision. This is because (1) all public actions affect distributional objectives that cannot be ignored, and (2) for many features of the environment, willingness to pay is either impossible to estimate or an inappropriate measure of the objective being served. For these reasons, economists evaluating public programs in the environment (and elsewhere) must be content with producing trade-off functions that will be evaluated subjectively by a decision maker. In a democratic society, the ultimate decision maker is presumably the public. In this view, then, it is not only legitimate, it is methodologically necessary that in matters of public policy the expert go "out on the stump" and present his results and interpretations to the public. This implies a more "activist" role than many are willing to accept, but it is the only way for the public to understand the implications of policies that affect them and to appraise the underlying values that determine both policy and analysis. In short, it is the only way to assure objectivity.

FOOTNOTES

[1]The National Park Service report that includes this letter is frequently called the Prewitt Report but is correctly referred to as "The Economics of Public Recreation: An Economic Study of the Monetary Evaluation of Recreation in the National Parks," National Park Service, United States Department of Interior, 1949, Washington, D.C.

[2]Secondary or indirect benefits are two terms referring to a class of economic events heretofore unmentioned in this book. These events can best be explained by example. When an irrigation project is developed, the newly developed area produces agricultural commodities. These commodities are called primary effects of the effort. In addition to these primary effects, there may also be increased sales of seeds, fertilizers and tractors. These sales (and hundreds of others like them) are described by economists as the secondary or indirect benefits of an irrigation project. Similarly, a whole set of costs (for roads, schools, sewage systems) called secondary or indirect costs arise when major investments are made. While economists are somewhat settled on definitions, the measurement and analysis of these items has always carried with it some vexing problems. Among these is the problem associated with a fully employed economy in which all secondary effects are merely transferred from other areas.

[3]Marion Clawson, *Methods of Measuring the Demand for and Values of Outdoor Recreation,* Reprint No. 10, (Washington, D.C.: Resources for the Future, Inc., 1959).

[4]Cost/effectiveness has not received the analytic attention that has been afforded benefit/cost analysis. This is likely only a problem of time. Some systematic treatment of the topic can be found in Stephen Enke, ed., *Defense Management* (Englewood Cliffs: Prentice-Hall, Inc., 1967) and in United States Congress Joint Economic Committee, "The Analysis and Evaluation of Public Expenditure: The PPB System" (Washington D.C., GPO), in three volumes, 1969.

[5]Herbert H. Stoevener and William G. Brown, "Analytical Issues in Demand Analysis for Outdoor Recreation," *Journal of Farm Economics,* vol. 50 (December, 1967), pp. 1295 - 1304.

[6]For a more detailed discussion of these and the following points, see David Seckler and L. M. Hartman, "On the Political Economy of Water Resource Evaluation," in David Seckler, ed., *California Water: A Study in Resource Management* (Berkeley: University of California Press, 1971).

chapter twelve

On The Strategy And Tactics Of Environmental Control

This chapter explores some of the strategic and tactical considerations involved in a program of environmental control. For expert advice on social - political action, one should read Saul Alinsky, the publications of the Sierra Club and Friends of the Earth, Martin Luther King, Jr., and the like. The following recommendations are biased — the "revolution" is not embraced. Revolution is necessary only in nonadaptive systems, and the basic structures of political and economic systems of the United States are taken to be reasonably adaptive. The concern of this chapter is the problem of how to capitalize on this adaptiveness.

STRATEGY

Any program of change must begin with the individual. While individualism is to be praised, it causes one of the more dismaying features of the "environment scene." Very briefly, current concern is expressed by innumerable well-intentioned individuals all launched on causes that usually have zero or even negative net effects. There is little question that the environmental movement can be regarded as a consumption good for many people. One can refrain from leaving trash scattered about, testify before the city council, write letters of deep concern to the *Times*, and even pick up others' trash on clean-up expeditions. Such activities alleviate guilt, provide a nice set of acquaintances, contribute to social standing, and are even fun. They also contribute in their own small way to the solution of the problem. But the way is all too small.

The problem of environmental control must be approached with all the calculation and determination of a businessman after his first million. It requires a cold assessment of the availability of time, money, and energy and their allocation in the direction of maximum effectiveness. The environmental problem is too important to be played as a witless game. The first strategic rule of environmental control is, therefore, *get serious.*

The second rule follows directly from the first. Economists have long recognized that division and specialization of labor is essential to increased productivity. For example, it is silly for an attorney to spend too much time on the picket line. And it is silly for the talented businessman concerned with the environment to give up his business to knock on doors. Rather, the businessman should make all the money he can then give it to the attorney and finance the pickets.[1] The attorney should spend his time bringing legal action against polluters. Pickets should not be required to raise money or compile legal briefs — they should be picketing. But in order to make the most of the division and specialization of labor, some organization must exist to direct and coordinate these many and diverse activities. The second strategic rule of environmental control is *get organized.*

Organizations can provide specialists — lawyers, propagandists, economists, pickets, and little old ladies in white tennis shoes — and combine these various talents and terrors into a truly effective force. They can provide offices, typewriters, printing presses, secretaries, and other fixed costs that must be spread out over a large number of people to reduce average costs. More important, they provide an integrated pressure group — a "class" rather than a mere "herd" — and political and economic systems are ultrasensitive to class movements.

The reason for this is that the reward and penalty configurations of both economic and political systems relate not to the total but to the margin. In the political system, for example, most candidates can depend upon a certain assured vote. Thus, the candidates of each party in an election are assured of at least *x* percent of the total vote because of party affiliation. What determines an election is not the 90 percent or so of the assured vote that the two parties will more or less evenly share but, rather, the uncommitted 10 percent. Most elections in this country are determined by an even smaller margin than ten percent. A candidate often wins by one percent or even one-quarter of one percent of the total vote. Such small percentages give politicians ulcers, but they also give power to an articulate, well-organized group of people that is disproportionate to its size. A block of 1000 committed votes presented to a politician is very important to him even in a mass of several million votes. It could swing the election, a fact the politician cannot ignore. Because of the predictability of the great majority of voters, there is a genuine tyranny of minorities.

This is particularly true when, in addition to providing votes, an organization can provide money and personnel to help in politicians' campaigns. For example, the elections committee of the AFL-CIO is a notoriously poor organizer of votes among union members. Usually, one half of the union vote cancels out the other half. Yet, the AFL-CIO enjoys large political power because it is able to provide its favorites with money, facilities, workers, and the organizational structures of local unions.

Much the same is true of business. Profits are made on the margin. About 95 percent of the total sales of a typical company merely covers operating and fixed costs. It is the sale of the last few units that brings profits and rewards to management.

Bureaucracies also respond to the margin. Indeed, they are more subject to minority rule than perhaps any other segment of the economy. Most people have never heard of the many regulatory agencies and public firms in federal, state, and local governments. As a result, many civil servants enjoy the immunity of anonymity from all but the few people who are determined to find out what they are doing. The people who most often do this are those who are regulated by the agencies. Thus, the pressure placed upon these officials comes from the highly organized, well-financed "minorities" they are supposed to regulate.

A well organized minority of environmentalists can, by exerting pressure on the margin of votes, sales, or influence, achieve results that would be impossible in a pure and informed democracy. Organizations, like individuals, should be specialized. If they do not specialize, they dissipate their energies in chaos and tend to paralyze into bureaucracies. Conversely, specialized organizations should coalesce to promote common interests. The great strength of the environmental movement in California lies in this: A multiplicity of specialized organizations have combined their various talents in alliances to promote matters of general concern. Thus, there are the Sierra Club, Friends of the Earth, the Planning and Conservation League, the Committee of Two Million, Ecology Action, the Audubon Society, and many others. But when a bill of common interest is before the legislature, these many organizations speak as one voice and often combine to finance lobbyists, research, and advertising.

Previous chapters frequently drew on the feedback concept. The discussion in this chapter to this point can be summarized in terms of a feedback concept but a peculiar feedback that amplifies each message it receives. The control or regulation of powerful machines (like an automobile or a bulldozer) is accomplished with the expenditure of effort vastly disproportional to the total energy of the machine. Through a series of elaborate mechanisms, the machine amplifies very minor "marginal" adjustments in the steering wheel, the accelerator, and the brake into large results. Political and economic mechanisms respond in similar fashion. They amplify marginal adjustments to the balance of power. A pictorial summary of social amplification on these lines is provided in Figure 1.

There is a third strategic consideration to discuss before going on to the examination of various instruments and tactics. This is more difficult to define, but it has to do with program selection. Every environmental organization faces the classic economic problem of allocating scarce means among alternative ends. The organizations have a bewildering variety of actions they could take; yet, if they do more of one, they must necessarily do less of the other. The most effective set of programs must be selected.

There are, of course, many elements involved in determining the most effective set, but the one explored here involves a distinction between two different kinds of programs. Some programs are concerned with structural changes in the system, while others seek to control a few crucial variables in the system. The distinction is elusive but important.

It has long been recognized that one can lose a battle and win the war. In choosing any particular battle — this or that dam, building, regulation, or development project — it is vitally important to assess not only the immediate costs and benefits of the act itself but also the long-run structural changes it entails. Even if it is known that the battle will be lost, one may rationally oppose a dam project. The opposition may make more people conscious of the environmental losses involved and this added opposition may weaken support for the next dam that is proposed. On the other hand, one can win the battle and lose the war. This has undoubtedly happened to some degree with the eruption of violence in racism and campus reform. A few improvements were made but only at the expense of massive reaction and alienation of general support. This principle is well-recognized in law. If an attorney can set a precedent for future cases, if he can introduce a structural change in the legal system, his success or failure regarding the particular case may be of little consequence. The third strategic rule of environmental control is, therefore, *strive for structural change.*

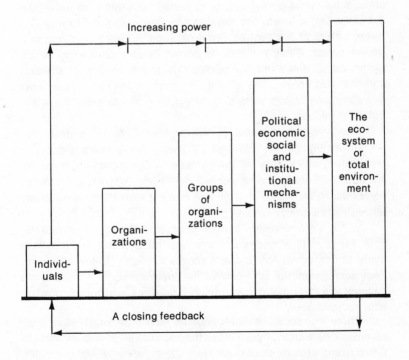

FIGURE 1: Increasing Power and Environmental Control

TACTICS

This section will briefly examine different kinds of tactics that either are being used or could be used in the fight for environmental control. The examples are designed to be suggestive only, for it is clear that each is a very large subject in itself.

1. The fundamental problem of control in cases of externalities and collective goods is the lack of clearly identifiable property rights. For example, rights to most amenities are very poorly defined; therefore, they are "up for grabs" to anyone with the power and opportunity to seize them. What is needed in cases of environmental degradation is an extension of the concept of private property into the domain of what is now collective property. In other words, an Environmental Bill of Rights with the full force of the law is needed to protect the public interest in collective goods.

Another important development in the law is recognition of class action suits by the courts. Generally speaking, an individual does not find it worth the time and money required to bring suit to the extent of his own personal damage in cases involving collective goods. Class action suits permit simultaneous legal action on behalf of many injured parties. Thus, the awards of the suit can be very high. The threat of large collective settlements through class action suits provides powerful disincentives to the potential polluter.

There is likely no area of environmental control quite so important as that of the law. Indeed, many environmental activists should first obtain a good background in environmental economics and then study law. A sophisticated and aggressive attorney can do more for social change and make it stick than perhaps any other professional.

2. High prices retard consumption, and low prices encourage it. A large part of today's environmental problem results from some prices being too low and others too high to accurately reflect social benefits and costs. The instruments of taxation and subsidy are effective devices for bringing market prices into line with social values.

Three commodities stand out as vastly underpriced. These commodities together constitute the root cause of a good part of the current environmental problem. They are electricity, water, and automotive transportation. The public has been convinced that "cheap and plentiful" supplies of electricity and water are essential to the prosperity, if not the very existence, of the nation. This philosophy has resulted in significant waste in consumption of these commodities and grandiose, environmentally destructive construction projects to provide them. Having exhausted the more conventional sources of electricity and water, the nation is being stampeded into the construction of nuclear reactors to generate electrical energy and to desalt seawater. Unless fusion reactors prove feasible, breeder reactors are the wave of the future. With breeder reactors, slight errors can be catastrophic. It is true that societies must, on occasion, risk catastrophic acts. Thus, World War II could have been a catastrophe, but it was believed that the issues justified it. One may be willing to risk catastrophe for democracy. But to risk massive irradiation so more aluminum beer cans can be made or so the housewife can flush more water and leave more lights burning or so a farmer can irrigate his fields in a slovenly fashion is surely questionable.

Steep progressive taxes should be placed on each additional kilowatt of electricity burned and each additional gallon of water used. Every effort should be made through such marginal taxation to destroy the rapidly spiraling demand for electricity and water. Attempting to meet, and even provoke, this rising demand through ever more elaborate production facilities and low prices appears perverse.

The same holds true for automobiles. Steep progressive taxes should be placed on the "pollution size" of automobiles. The pollution size index may be emissions of pollutants, horsepower, weight, or some combination of these standards. Thus, one of the typical American automotive behemoths might bear a tax equal to or even greater than its market price while small, more pollution-free cars should pay no, or very little, tax. Substantial reduction of the numbers and size of automobiles in this nation would perhaps contribute more to solving the problems of the city than any other device.

If such programs of taxation were carried out, large revenues could be obtained to subsidize other more socially beneficial activities. An obvious choice would be an adequate system of public transportation. Most Americans have never experienced the pleasure of an efficient public transportation system and the benefits it offers to the pocketbook.[2] The absurdity of the automobile culture is apparent to those that have. It may be necessary to run such a system by electricity but that is all the more reason to price electricity out of wasteful uses and use tax proceeds to encourage more beneficial alternatives.

3. The previous tactics are "macro" in nature — they require a change in social policy by governments. "Micro" tactics that can be carried out by only a few dedicated individuals can also be very productive. Such is the case of the "Pelt War."

Some months ago, a few people decided to campaign against the use of pelts of endangered species — tigers, seals, and alligators, among others — as items of ornamentation (conspicuous consumption). Because of the shift in social standards brought about by this campaign, it became unfashionable to parade about in these skins. The effect has been the virtual destruction of the domestic market for these commodities. Even the international furriers associations are now refusing to traffic in the trade of endangered species.

There appears to be a rising awareness of the absurdity of conspicuous consumption, especially among the young. This movement could eliminate much of the demand for status symbols. One can now drive a Volkswagen and retain one's standing. The next great advance would require that one *lose* status on being seen driving a Cadillac. A severe collapse in status demands would provide a basis for resolution of environmental problems by destroying the demand for an ever spiraling GNP. Stop buying.

4. Another important model for environmental control has been provided by Cesar Chavez and his United Farm Workers Organizing Committee (UFWOC). This is the boycott — refusal to purchase the product of certain firms.

A word of background is in order. The grape and lettuce industries have been acquired over the course of time by large, national, nonagricultural firms, with many different product lines whose importance far exceeds that of their agricultural activities. For example, the Purex Company is one of the largest lettuce growers, but its lettuce business is trivial in comparison to its position in the soap and cleanser industry. Also, supermarkets sell a variety of commodities among which grapes and lettuce are trivial items. Both the production and distribution of grapes and lettuce are dominated by firms from whose lofty view these are very marginal products. Therein lies the secret of UFWOC and the genius of Cesar Chavez.

Briefly, a boycott was declared against nonunion grapes *and* those who sell nonunion grapes. As a consequence, not only the supermarket sales of grapes declined, but total sales of all commodities fell as well. The sale of grapes in a supermarket is trivial; the sale of everything else is not. Supermarkets found the tail of grapes wagging the dog of total sales, and this created an intolerable situation. Also, certain disruptive tactics of peaceful picketing and sit-ins were introduced to disturb the distribution and retailing systems. Large systems depend on fine-tuning and lock-step integration of functions. The slightest variance amplifies through the system to produce chaos. The net effect was well summed by the manager of a West Coast chain: "A year ago, I never thought of a grape; now my executives and I think of little else." UFWOC won.

In sum, Chavez created a "nuisance," and the nuisance simply was not worth it. Environmental activists can employ the same tactics against firms that refuse to sell returnable bottles or that sell environmentally destructive detergents or that refuse to treat pollution emissions from their factories.

5. The stock exchanges of this country provide markets for investors in public corporations. There is nothing so vital to the investor as the change in the values of his stocks. For this reason, there is no fact so vital to the managers of these corporations as the change in value of their company's stock. The stock exchange is at once one of the most powerful institutions of society and one of the most adaptable. It is natural to ask whether this power and adaptability cannot be used to regulate environmental quality.

A good part of the fluctuations in the values of stocks are due to the influence of portfolio managers — that is, managers of mutual funds, large estates, foundations, investment companies, and the like. The manager has to decide every day with respect to every stock whether to buy or sell, depending upon whether he expects the price of the stock to go up or down.

Imagine that an "Investors Against Pollution" organization is established. This organization undertakes to discover in a conscientious way the five greatest polluters whose stock is traded on the exchange. They would also find those five companies whose efforts in pollution control were most commendable. Having established these facts, they would publish in the *Wall Street Journal* the names and specific grievances against, or commendations to, these companies. They would recommend either explicitly or by implication that all right-thinking investors should sell the "polluters" and buy the "cleans."

Two consequences could be expected to follow from such an advertisement. First, there are a great many people who own stocks and who have a strong commitment to the environment. Moreover, the typical investor is not highly committed to particular stocks. So the first effect would be that a good many people across the country would sell their "polluter" holdings and, assuming other factors were favorable, buy into "cleans." It is a well-recognized fact that the environmental issue is a concern of the middle- and upper-income classes and that it represents a great potential of economic power.

Second, there is the case of the cold, calculating portfolio manager who may not give a damn about the environment. The portfolio manager typically comes to work in the morning and, after some study, compiles two lists of stocks — those he thinks will go up and those he thinks will go down. Once he has made these lists, he has to decide which stock in each particular list to buy or sell, for he usually cannot change a position in all of them. Clearly, he will buy or sell those stocks he thinks will go up or down the most. But he often has no certain way of knowing, so he chooses a particular stock by a more or less random process involving hunches and intuition.

He has just read the *Wall Street Journal* and has seen the "Investors Against Pollution" ad. "What are those silly *!%&! up to?" he says to himself. "Some people may be stupid (read: moral) enough to follow the ad!" He quickly runs through his list of buys and sells. He finds the following: In his list of buys is a "clean," so he buys in anticipation that the price will rise even more than he thought because of purchases made by environmentalists. But there is also a "polluter," so he decides not to buy or to buy less. The same process works in reverse in his list of sells. The portfolio manager himself does not care about pollution but, if he thinks a significant number of stock owners will care, he will *behave just as if he cared* in order to make money. Because of the response of professional managers and speculators, the advertisement could work very well even if only a few respond to it because of a moral commitment to the environment.

In this way it is conceivable that, working through the mechanism of the stock exchange, one could effect capital gains and losses on the order of millions of dollars to firms who invest some thousands of dollars in pollution control or who refuse to do so. It is not necessary to follow the repercussions of such a campaign through the stockholders, the board room, and the management of the affected companies; but it would be surprising if a tactic of this sort did not change some company's policies with respect to the environment.

6. Finally, there is the most powerful tactic of all, the ultimate structural change, *to change what people think.* Compared to the long-run influence of the great intellectuals of history, the short-run influence of politicians and warriors is rather small beer. As Keynes, one of the greatest, observed:

> Practical men, who believe themselves to be quite exempt from any intellectual influences, are usually the slaves of some defunct Economist. Madmen in authority, who hear voices in the air, are distilling their frenzy from some academic scribbler of a few years back. . . . soon or late, it is ideas, not vested interests, which are dangerous for good or evil.[3]

FOOTNOTES

[1]For example, Alvin Duskin, the well-known San Francisco dress manufacturer, uses his business to finance environmental work and social change. Gordon Sherman, president of Midas Muffler, has used his large fortune to similar ends. Sherman put the essential point perfectly when he declared that his ideal is to use the oil depletion allowance to avoid taxes and then use that money to lobby against the oil depletion allowance (Nicholas Von Hoffman, Times - Post Service, in the *San Francisco Chronicle,* March 1, 1970). According to *Life* magazine (January 22, 1971), Howard Hughes is prepared to dedicate his life and every cent he has to stopping nuclear testing of all kinds.

[2]For example, the systems in Montreal, London, Paris, and Moscow.

[3]J. M. Keynes, *The General Theory of Employment, Interest, and Money* (New York: Harcourt Brace Jovanovich, Inc., 1935), pp. 383 and 384.

chapter thirteen

Economic Growth
And
Environmental
Decay

For a considerable number of pages, problems caused by the intermixing of economic and ecological systems have been examined. While economics has a substantial contribution to make to solving these problems, it is certainly not a sufficient basis of analysis — much less a coherent synthesis of all these complexities. As was mentioned at the outset, a study of environmental problems is the study of unintended consequences of choice. Although the *unintended* aspects indeed remain, they need not be *unrecognized*. Recognition of side effects, evaluation of trade-offs or compromises, and analysis of alternative paths of action are the skills that the economist can bring to the study of contemporary problems confronting man and nature.

The important points of some previous chapters need re-emphasis. The first part of this book was devoted to themes that ordinarily appear under the heading "economic growth." In sum, those chapters said that growth is not always what one thinks it to be. The most common measure of "growth" is so defined as to include, in a positive way, many things that actually are detrimental to health and happiness. Indeed, GNP does measure the output of an economy. But collecting garbage, the costs of driving to get away from congestion, and the cost of cleaning up polluted streams all enter the GNP calculation in the same way as does the production of woolen underwear. An economy seeking growth (as most economies do) and measuring growth with reference to GNP is likely to be highly disillusioned at the outcome of its efforts. At the very least, a reformulation of the national accounting system is in order. A system that places "goods" in one account and "bads" in another would be much superior to the present system. One suspects, though, that changing the accounting system would get more things out for public inspection but would not touch the fundamental problems.

All organized economies seek to grow. This has been true since the dawn of civilization and has been especially true since the Industrial Revolution gave humanity a reprieve (or at least a temporary stay) from the centuries-old struggle between man and nature. Some economies have done much better than others. The United States economy has flourished as has no other, probably because of a unique combination of technical skill, natural resource availability, and institutional arrangements. In 200 years, this combination has allowed the citizens of the United States to develop a productive capacity so refined as to be able to provide six percent of the world's population with 35 percent of the world's produced goods and services. This is said without apology, but at the same time some rather awesome questions must be raised. Can growth be maintained? If it is maintained, what will be the cost in terms of resources and environmental quality? How will incomes be distributed in this nation and among other nations in the global community? Should a few Western economies continue to grow, while many economies in the world continue to lie dormant or at best grow at almost imperceptible rates? These questions (and many others) are fundamental to the future of humanity. The student of environmental economics, by himself, cannot provide unequivocal responses to any of these; he can provide honest lists of alternative responses to each of them. His value lies not in the answer itself but in the direction he can give to the decision-making process.

The message of the second part of the book is very simple. *If* all externalities were to be properly accounted for through recognition and compensation; *if* all collective goods were produced and distributed in an optimal fashion; *if* future generations were cared for with reference to their tastes, preferences, and the technology available to them; *if* there were a satisfactory distribution of income through society; if . . .; if . . .; if . . ., *then* the market could be relied upon to provide an optimal rate of growth. Market prices would then be "rational" in the sense that they would adequately reflect all social benefits and costs. Valid decisions for the present and the future could be made based on this information. This is not to say that the environment would remain unscathed under such an ideal system. But if degradation were present, it would be occurring in a recognized fashion, and the relevant costs and benefits would have been taken into account. (Economists have recognized this set of requirements as the "Pareto-optimality conditions.")

To think all these conditions and circumstances can be met and accounted for is overly ambitious. The requirements for an ideal economic system are reasonably well known, but the mechanisms for meeting these requirements are poorly understood. Since the real world does not correspond to the ideal, the process of growth becomes much more ominous, bringing with it those oft-mentioned, unintended aspects — externalities, loss of collective goods, and the potential impoverishment of future generations. Insofar as these imperfections exist, market prices will not accurately signal values; and insofar as growth continues under these irrational prices, society is led away from its desired state. The growth process itself becomes irrational.

Inevitably, one is led back to the basic question: Why grow at all? This economic society has the productive capacity to comfortably satisfy the basic requirements of all of its constituents. If added growth is recognized as a threat to the environment and to the planet, do the benefits warrant the costs? This view is not likely to be a popular one since everyone's life is impregnated with strong forces toward growth almost from the time of birth. Some reconsideration must be made though. The reconsideration has two major components. First, it must be determined that it is possible to slow down. Second, the likely effects of low or zero economic growth must be considered.

With respect to the first question, it must be realized that the rapid rate of economic growth in the United States over the past few years is by no means a "natural" phenomenon. It is the product of calculated choice by high officials of the United States government.[1] This policy has been encouraged and implemented by the Council of Economic Advisors and the monetary authorities of the Treasury and the Federal Reserve systems. High rates of economic growth are a consequence of advances in modern economic theory stemming from J. M. Keynes' great work, *The General Theory of Employment, Interest, and Money,* first published in 1936. Before these advances in economic theory, economic growth was virtually a natural phenomenon. It was beyond the range of choice because people did not know how to regulate it. Now, governing the rate of growth is comparable to adjusting the carburetor of an automobile. Like all inventions, economic theory only says "how to do it," but its power can be abused. It is no exaggeration to say that many environmental problems today are the external costs of using modern economic theory, just as air pollution results from excessive use of the automobile.

Modern theory shows that the growth rate of the economy can be controlled through injections and withdrawals of money in the system (by balancing and unbalancing governmental budgets) and by a variety of monetary devices. These basic ideas have become increasingly sophisticated so that economists now know much more about how to make the injections and withdrawals of money, where to make them, and when to make them.

Clearly, society can pick and choose its rate of economic growth from a very large range of alternatives. For example, in the past two years one of the external costs of growth has forced society to temporarily abandon the goal of rapid economic growth. Inflation reached an annual rate of nearly six percent in 1970. In alarm, the powers that control the performance of the economy reduced the growth rate to zero. This caused unemployment to rise to a rate of 5.6 percent per year. This is a good case of the treatment possibly being worse than the illness. The point is, again, that the growth rate can be chosen. And society can choose a zero growth rate (no growth rate at all) if it wishes.

This leads to the second question: What would be the likely effects of a slow, or zero, rate of economic growth? There would be many, but the first would likely be rising unemployment. If, in the interest of environmental quality, society chooses a low growth rate, would this not inevitably cause massive unemployment? The answer is both yes and no. It is "yes" if the prevailing structure of employment opportunities continues; the answer is "not necessarily" if this structure were revised.

Labor productivity is expected to rise during the next ten years at an average annual rate of two percent, and the labor force will grow over this decade at approximately the same rate. Under the existing structure, then, the economy must grow at an average annual rate of four percent, just to keep people employed. If the present disposition toward work is to be maintained, then we have little choice. We must grow because we must work. If, however, people were willing to work less, more people could work. If labor would reduce its work week by four percent per year, full employment could be maintained at a zero rate of growth. Of course, per worker income would decline by about two percent per year because of increased numbers of workers, but once the labor force reached a stable point, full employment could be maintained at growth rates of from zero to two percent; leisure time and/or increased wages would absorb the increased productivity.2

It is interesting to speculate on the impact of increased leisure on the environment. At first, it may seem that more leisure would have a detrimental effect because more people would be using the already crowded recreational areas. This is not necessarily so. Assume that American industry and society adopted a three-and-one-half-day work week. Further, assume that working time was staggered through the seven days of the week. This would introduce a profound qualitative change in the use of leisure. The current "weekend" is too short to be used for some purposes, but too long to be merely wasted. Given the three-and-one-half-day work week, a good part of the population could have second homes in recreational areas within reasonable distances from their work. These homes could be cooperative ventures with a friend who worked the "other" days of the week. Two effects would immediately follow: (1) the average level of congestion in the city would diminish substantially since many families would, at any given time, be away; and (2) the weekend congestion problems that afflict recreational areas would be substantially relieved. Use of these areas would be averaged out over all the days in the week. "Peak load" problems associated with employment time would be alleviated and fixed capital in offices and factories better utilized through the seven-day period.3

Finally, there is some evidence that, with the noticeable decline in the levels of conspicuous consumption of the past few years, people would freely elect reduced work weeks if given the opportunity.4 Without conspicuous consumption, one does not need to work as much to be well off. With sufficient leisure time, the costs of living are less because one can build one's own things, grow some part of one's food, and spend less on escapist activities. If a market for leisure were offered in a sensible way through adaptations in the structure of employment, many would "purchase" more of this good through working less.

Increased leisure could also help resolve the problem of the inequality of income. If enough affluent workers elected to use more leisure time, a strong, effective demand would be created to train and employ people who are now denied access to the labor market because of their background and training. The unemployed and underemployed are now in a state of enforced leisure just as some of their contemporaries are in a state of enforced work. Each has too much of what the other needs, so reallocations on the margin should be made.5

In a near stationary state, the amount of investment required in the economy would be reduced to levels sufficient only to replace the wear and tear on capital equipment — machines, houses, roads — and to provide for technological advance. Since the investment would be less, fewer savings would be needed. The high "saving capacity" of the wealthy would be much harder to justify, as would the accumulation of fortunes through inheritance or other means.

Now, it is not desirable either to overstate the advantages of a stationary state or to minimize some of the real sacrifices that would be entailed. Among the latter, certainly, is the question of impact on national defense and international trade relations. But such problems cannot be solved here.6 Rather, the intent has been to point out the necessary connections between economic growth and environmental quality. The more society has of one, the less it can have of the other. This has been the basic thesis of the book. From this thesis flow three direct consequences that cause significant departures from the mainstream of thought in ecology, economics, and the environmental movement.

First, ecologists have sometimes overstated the problem in their pronouncements of physical catastrophe. Far before the physical apocalypse arrives, the battle of human values will be irrevocably lost. The life style of Perelman's "terminatory" is a far more immediate threat than the "termination" suggested by some ecologists.

Economists have underestimated the problem in believing that the detrimental effects of economic growth in the environment are simply by-products, spillovers, or external costs of growth. This leaves the impression that, if society uses its conventional powers over economic policy to "internalize" these externalities, the process of economic growth can continue *ad infinitum* in a pleasant environmental setting. This is a wholly misleading belief. Environmental decay is an integral part of the economic growth process. There is no conceivable way to produce, use, and eventually dispose of economic commodities without creating some degree of environmental decay. The longer economic growth proceeds, the more severe will be environmental decay. It is a trade-off. The trade-off is acceptable to a point, but beyond this point it can no longer be tolerated. To believe that society can always have both is simply a reaffirmation of that ancient fallacy of "something for nothing."

Third, it is not sufficient for affluent people, who compose a good part of the environmental movement, to be willing to sacrifice further increases in the conventionally defined standard of living in order to restore and maintain the environment. They must also be willing to raise the poor to above the poverty line and to ensure that people who lose their jobs through reductions in employment opportunities still have an adequate income. Every environmental problem, from the SST to the local dam, is also an employment problem. The affluent must be willing to share with the poor and unemployed.

In sum, in order to restore and maintain a high level of environmental quality, it is necessary to accept low rates of economic growth; and in order to ensure that everyone benefits by such a policy, it is necessary to provide a minimum income to people whether they work or not.

The book began with the early economists; it is well to close with one of the greatest. John Stuart Mill believed, as did his predecessors, in the inevitability of the stationary state. In Book IV of his *Principles of Political Economy,* he says:

> . . . that at the end of what they [Mill's fellow economists] term the progressive state lies the stationary state, that all progress in wealth is but a postponement of this, and that each step in advance is an approach to it. . . . The richest and most prosperous countries would very soon attain the stationary state, if no further improvements were made in production arts, and if

there were a suspension of the overflow of capital from those countries into the uncultivated or ill-cultivated regions of the earth.[7]

In so saying, he echoed the inevitability of the stationary state propounded by his predecessors. But, unlike his predecessors, Mill looked somewhat joyously upon the prospect of the stationary state:

> I cannot . . . regard the stationary state of capital and wealth with the unaffected aversion so generally manifested towards it by political economists of the old school. I am inclined to believe that it would be, on the whole, a considerable improvement on our present condition.[8]

Mill's reasons for his views center primarily on the release of men from the struggle to get ahead of their fellows. Men would be better able to enjoy life. There would be some work, but there would be no amassing of huge fortunes, only modest inheritances and an increasing amount of leisure time to devote to more rewarding pursuits:

> This condition of society, so greatly preferable to the present, is not only perfectly compatible with the stationary state, but, it would seem, more naturally allied with that state than with any other.[9]

There is something compelling about the conviction that the stationary state offers great benefits to a society. The early Classicals primarily feared the stationary state; Mill welcomed it. The time has come when Mill's thinking should be reexamined. The stationary state offers the bounty of both man and nature. The great dread of the early Classicals may be the only hope of the Moderns.

FOOTNOTES

[1]For an excellent statement of the policies and preconceptions of modern growthmanship by the growthmen, themselves, see Walter Heller, *Perspectives on Economic Growth* (New York, Random House, 1968).

[2]Productivity would change under such circumstances. It could tend to go down because of less "embodied" technological

change in capital equipment. On the other hand, it may tend to rise because of a well-rested and more enthusiastic labor force. Many alternative models of a zero growth rate economy could be generated. For example, if investment in machinery were curbed sufficiently, everyone could work in a "Chinese laundry" economy. But this is a rather absurd solution. Contrary to much popular opinion, the goal of economic activity is not employment or work — it is increased human welfare.

[3]We are not so naive as to think that these changes would come easily. Drastic changes in habits of education, worship, and holi- to reestablish these institutions on a seven-day basis.

[4]In the first part of 1970, fully two-thirds of all new car sales were in the $3100 and under class. As the manager of the Chevrolet Division of General Motors says, "It's the American system that's being tested." In *Fortune* (October, 1970), p. 36. See also Judson Gooding, "Blue Collar Blues on the Assembly Line," *Fortune,* (July, 1970) and *Time,* (November 9, 1970), pp. 69 - 72. Another innovation is the 40 hour, four-day week recounted in Riva Poor *4 Days — 40 Hours* (Cambridge: Bursk and Poor, 1970). Counter-arguments are provided in Steffan B. Linder, *The Harried Leisure Class* (New York: Columbia University Press, 1970), and Gilbert Burck, "There'll Be Less Leisure Than You Think," *Fortune,* (March, 1970).

[5]It may also be noted that the advance of mechanization (auto-mation) will likely make it impossible to maintain high rates of employment, even under rapid growth. Thus, it is likely that the unemployment problem will have to be confronted whether growth occurs or not.

Another barbaric aspect of this employment problem is the policy of controlling inflation through forcing unemployment. The unemployed are forced to pay the costs of the inflation that growthmanship imposes on society. They should be compensated through high unemployment payments. Also in recession periods everyone should be forced to work less through reduced work weeks rather than permitting some to work the same and forcing others not to work at all, as is now the case.

[6]E. J. Mishan has addressed several of these problems. "The Costs of Economic Growth," *Op. cit.,* Parts 1 and 4.

[7]John Stuart Mill, *Principles of Political Economy* (London: Long-men's Green and Company, 1909), p. 746.

[8]*Ibid.,* p. 748.

[9]*Ibid.,* p. 750.

B 3
C 4
D 5
E 6
F 7
G 8
H 9
I 0